Reactivity and Structure
Concepts in Organic Chemistry

Volume 15

Anthony J. Kirby

The Anomeric Effect and Related Stereoelectronic Effects at Oxygen

With 20 Figures and 24 Tables

Springer-Verlag
Berlin Heidelberg New York 1983

Anthony John Kirby
University Chemical Laboratory, University of Cambridge
Lensfield Road
Cambridge CB2 1EW, U.K.

List of Editors

ISBN 3-540-11684-2 Springer-Verlag Berlin Heidelberg New York
ISBN 0-387-11684-2 Springer-Verlag New York Heidelberg Berlin

Library of Congress Cataloging in Publication Data. Kirby, Anthony John. The anomeric effect and related stereoelectronic effects at oxygen. (Reactivity and structure; v. 15)
Bibliography: p. Includes index. 1. Isomerism. 2. Stereochemistry. 3. Oxygen. 4. Chemical bonds. I. Title. II. Series.
QD471.K5 1982 547.1′2252 82-5889 AACR2

Printing and binding: Konrad Triltsch, Würzburg
2152/3020-543210

Preface

The anomeric effect is a well-defined and much studied example of a deep-seated and very general effect in organic chemistry. It involves a fundamental interaction between the lone pair electrons of oxygen and the antibonding orbitals of adjacent bonds, which can control not only conformation but also reactivity in suitable systems. So the heart of this book is a consideration of the properties of non-bonding pairs of electrons, and the influence of orbital-orbital interactions on structure and reactivity.

I have been lucky in recent months to be able to discuss various matters arising in this book with people who know more about them than I do. Particularly Professors C. Altona, A. Eschenmoser, F. Hirshfeld, M. Lahav, R.A. McClelland, G.M. Sheldrick and A. Vos, and, nearer home, I. Fleming, P.G. Jones, A.J. Stone (who helped with the rather theoretical Appendix) and I.H. Williams. I am also most grateful to Professors P. Deslongchamps, J.D. Dunitz, D.G. Gorenstein, W.P. Jencks, G.A. Jeffrey, R.U. Lemieux and S. Wolfe, for very helpful correspondence, and information in advance of publication.

I have covered the literature received in Cambridge up to early December 1981.

Cambridge, June 1982 — Anthony J. Kirby

Table of Contents

Introduction 1

A. Stereoelectronic Effects on Structure 3

1. Introduction 3
2. The Anomeric Effect 4
3. The Reverse Anomeric Effect 15
4. The Exo-Anomeric Effect 18
5. The Generalised Anomeric Effect 20
   a. Sulphur Compounds 23
   b. Transmission through Phosphorus 26
   c. Transmission through Sulphur 31
6. The Gauche Effect 32

B. The Electronic Basis of the Effects 37

1. Introduction 37
2. The Status of Lone Pair Electrons 39
   a. Evidence from Hydrogen-bonding 42
   b. Electron Density Measurements 44
3. Calculations on X⌒Y Systems 48
4. Effects on Bond Lengths and Angles 52
5. Effects in Furanoses 61
6. The Stereoelectronic Basis of the Anomeric Effect 62
   a. The Structure of CH_2F_2. 62
   b. Bonding in $ROCH_2X$ 64
   c. Bonding and Restricted Rotation in Dimethoxymethane . . 67
   d. The Generalised Anomeric Effect 71
   e. The Reverse Anomeric Effect 71
   f. The Exo-Anomeric Effect 72
7. Related Stereoelectronic Effects 72
   a. The Anomeric Effect at sp^2 centres 72
   b. The Gauche Effect 75

C. Stereoelectronic Effects on Reactivity: The Kinetic Anomeric Effect 78

1. Introduction 78
   a. The Use of sp^3-hybrid Lone Pairs 78
   b. The Stereoelectronic Basis of the Kinetic Anomeric Effect. . 79
2. Reactions of the Acetal Group 84
   a. Acetal Hydrolysis 84
   b. Acetal Formation 94
   c. Acetal Oxidation 96
3. The Hydrolysis of Orthoesters 102
   a. Hemiorthoesters 105
4. Carbonyl-Oxygen Exchange Reactions of Esters and Amides . . 112
5. The Hydrolysis of Imidate Esters: Stereoelectronic Effects at Oxygen and Nitrogen 117
6. Reactions of Phosphate Esters 126
7. Radical Reactions 128
8. Enzyme Reactions 132

D. Appendix 135

1. Canonical Lone Pair Description 135
2. Hybrid Lone Pair Description 136

E. References 137

F. Subject Index 147

Introduction

A molecule's bonding electrons serve not only as its skeleton, but also as a rudimentary nervous system. Perturbations (stimuli) at one position can be transmitted through the bonds to affect other centres more or less efficiently, depending on the distance and the local geometry. And conformation-dependent orbital-orbital interactions can have important effects on the stabilities of both ground and transition states. Evidently we must learn to understand such fundamental interactions if we are to understand how molecules behave, and especially how they react.

There is a growing awareness of the importance in this context of the stabilising two-electron interactions possible between antibonding orbitals and filled, bonding or non-bonding orbitals. Interactions of this type are strongest when they involve non-bonding electron pairs, and give rise to a number of readily observed effects on the conformations of suitable systems, particularly systems containing bicovalent oxygen.

The most important of these is the anomeric effect, first identified as an anomalous effect on sugar conformations, but now recognised as being of much more general importance, and related to a number of other factors which influence conformational equilibria, most notably the gauche effect. Furthermore, it is becoming clear that the interactions which give rise to these stereoelectronic effects on conformation may have similar, and potentially larger, effects on reactivity. The 'kinetic anomeric effect', in particular, is the basis of Deslongchamps' theory of stereoelectronic control, which has made organic chemists much more aware of the way the properties of non-bonding electron pairs can control reaction pathways.

This book describes our present state of knowledge of a range of effects of this sort, and traces them to a common stereoelectronic origin. The discussion is in three sections, which build on each other but are intended to be otherwise independent. Section A describes the anomeric effect, and related effects on conformation, including the gauche effect, purely phenomenologically. Explanations are not attempted, and lone pairs are scarcely mentioned except in the final chapter. Section B, on the other hand, is largely about non-bonding electrons. It begins with a description of the current status of lone pairs, as revealed by the latest physical and theoretical methods, and then goes on to discuss the stereoelectronic basis for all the effects discussed in section A, in terms of orbital-orbital interactions. Finally, section C discusses both the experimental evidence and the theoretical basis for stereoelectronic effects on reactivity.

In Sections A and C, which are mostly organic chemistry, I have used the familiar representation of the non-bonding electron pairs of bicovalent oxygen in terms of two equivalent localised sp^3-hybrid orbitals. The more theoretical discussion in section B uses the formally more correct description in terms of two non-equivalent canonical orbitals, familiar to theoretical chemists. In fact a careful comparison of these alternative representations in section C1(a) shows that for most purposes they can be considered as equivalent.

A. Stereoelectronic Effects on Structure

1. Introduction

The ideas of conformational analysis [1] are fundamental to a proper understanding of the structure and reactivity of organic compounds. Students are introduced at an early stage to the idea that rotation about single bonds is not free, but restricted by energy barriers which depend on the substituents at each end of the bond, and the dihedral angles between them. It is not necessary to understand in detail the quantum-mechanical basis for the interactions concerned to appreciate that the favoured conformation about the central bond of *n*-butane, for example, is staggered, with the terminal methyl groups *antiperiplanar* (*1*) to each other. Evidently the (unfavourable) interactions between the largest groups dominate in this simple case, and are minimised when these groups are rotated as far away from each other as possible.

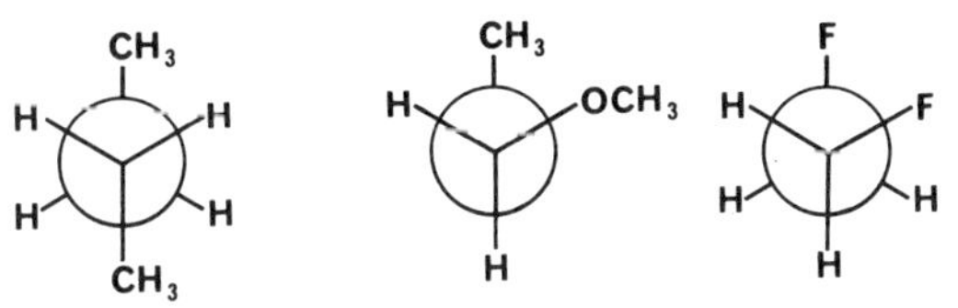

1 *2* *3*

This Section is concerned with systems which break this simple rule. The favoured conformation about either of the C−O bonds of dimethoxymethane $(MeO)_2CH_2$, is not *antiperiplanar* but *synclinal* (or gauche)[1] (2; see below pages 20–23): so too is that (*3*) of difluoroethane [2]. The preferences involved are not large in terms of free energies, but they are evidently large enough to dominate the 'usual' conformational factors in these systems. And since conformational preferences can have major effects on reactivity, it is clearly essential to identify the special factors involved if we are to understand the chemistry of such systems.

The compounds concerned are principally those (*4*) with at least two electronegative substituents on the same carbon atom, though the analysis

---

[1] Current usage prefers *antiperiplanar* to the less precise *anti* or *trans* to describe conformation *1*, but has been less decisive in replacing gauche by the equivalent and more systematic *synclinal* to describe conformations *2* and *3* (dihedral angles of 60°).

4

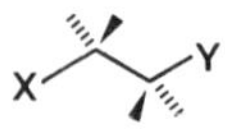

5

presented below (see page 75) is general enough to include some monosubstituted alkanes and vicinally disubstituted compounds (*5*). Structure, bonding and reactivity in such systems all show special features which have presented substantial difficulties of interpretation until quite recently. Thus the C−X bonds of CX_nH_{4-n} become progressively shorter as the number of heteroatoms bound to the central carbon is increased: and the reactivity of a molecule like dichloromethane is remarkably different from that of methyl chloride. Six-membered rings in some cases prefer the chair form with all substituents *axial*, or may even go over to the twist boat conformation. And these effects on ground states may be magnified when operating on transition states, thus having important effects on reactivity.

The most striking effects are observed when one or both of the heteroatoms of (*4*) is oxygen, and relate to the conformation about the C−O bond; which in turn has remarkable effects on the structure and reactivity of such compounds as acetals and orthoesters and their derivatives. The initial observations were made by carbohydrate chemists, and concerned the configuration at the anomeric centre of pyranose sugars. We will look first at the evidence for the major effects which have been identified, before going on to consider explanations and more detailed structural evidence.

2. The Anomeric Effect

The introduction of a single oxygen atom into a carbon chain or ring has only a minor effect on its conformation. The preferred rotational isomer of methyl ethyl ether (*6*) is *antiperiplanar*, like that of *n*-butane, though the conformatio-

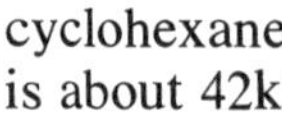

6

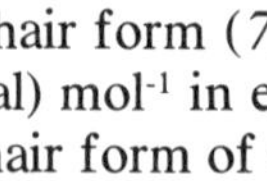

7

nal preference is significantly stronger in the case of the ether [3]. The conformational properties of the tetrahydropyran ring similarly resemble those of cyclohexane. The chair form (*7*) is preferred [4], and the barrier to inversion is about 42kJ (10 kcal) mol^{-1} in each case [5–6].

The preferred chair form of a substituted cyclohexane is generally the one with the larger number of equatorial substituents, and thus the minimum number of gauche interactions and non-bonded repulsions between axial substituents [1]. Similar considerations apply to substituted tetrahydropyrans. As for methyl ethyl ether *vis à vis n*-butane there are quantitative differences, because replacing a methylene group by an oxygen atom means that one axial H is missing, so that some diaxial interactions are reduced. And gauche and 1,3-diaxial interactions across oxygen, as in (*8*), are increased because C–O bonds are shorter than C–C bonds [5].

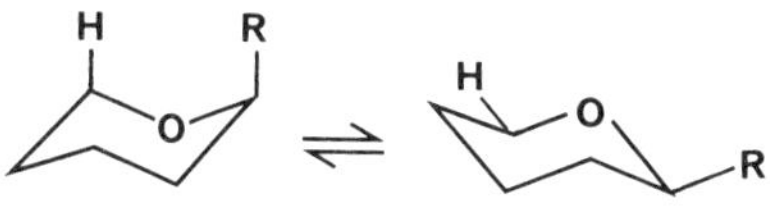

8

Nevertheless the general principles of conformational analysis apply to tetrahydropyran and its derivatives just as they do to carbocyclic systems, and their application [7–10] to the structures of the pyranose sugars marked a major advance in carbohydrate chemistry. To take one very simple example, there is an obvious correlation between the natural abundance of sugars and their conformational stabilities; so that the favoured β-anomer of D-glucose, the most abundant of the hexoses, turns out to have the uniquely stable structure (*9*) with all its substituents equatorial.

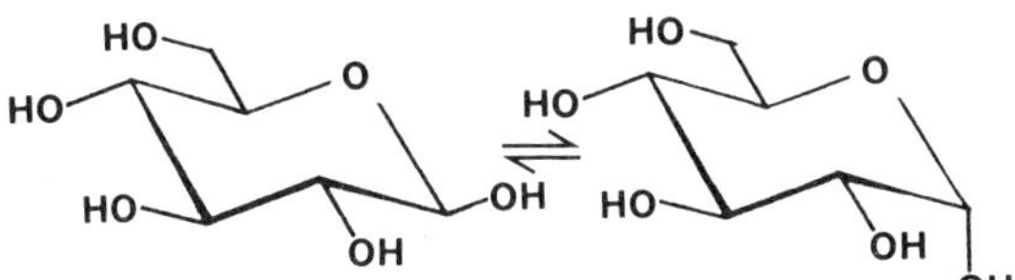

9 *10*

So far so good. But as soon as one attempts to estimate the position of conformational equilibrium for almost any pyranose derivative, using the methods so successful in the conformational analysis of substituted cyclohexanes, it becomes apparent that some additional and quite new factor is at work. Such calculations involve estimating the energies associated with the usual non-bonded interactions between substituents – gauche interactions, 1,3-diaxial interactions between OH and OH, OH and H, and so on – and summing them for the two possible chair conformations. Details, and examples of such calculations, are available in the secondary literature [8–10], but the anomeric equilibrium between α and β-D-glucose provides a simple illustration These isomers (*9* and *10*) differ only in the configuration at the anomeric centre, so only interactions involving the anomeric hydroxyl group are relevant to the position of conformational equilibrium. In both isomers this group has a gauche interaction with the hydroxyl group at the 2-position, so the only apparent difference is the presence of two 1,3-diaxial (OH : H) interactions in the α-isomer (*10*). These are the same interactions responsible for the equatorial preference of the hydroxyl group of cyclohexanol, so the position of conformational equilibrium between (*9*) and (*10*) should be similar to that between (*11*) and (*12*) under the same conditions.
The free energy difference between (*11*) and (*12*) defines the 'A-value' [1a] for the hydroxyl group. This is significantly greater in aqueous solution than

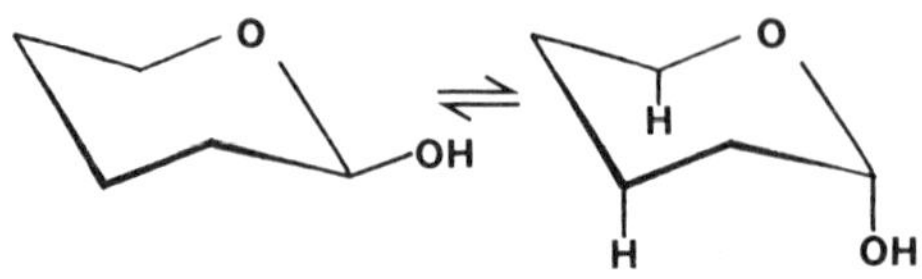

11 12

in aprotic solvents, probably because H-bonding solvation increases the effective size of the group, and a value of 1.25 kcal mol^{-1} has been suggested by Durette and Horton [10a]. This figure predicts an equilibrium mixture containing 89% equatorial (*9* or *11*) and 11% axial (*10* or *12*) isomer. In fact there is 64% of β-D-glucose (*9*) and 36% of the axial anomer (*10*) present at equilibrium in water at 25° [9a]. This difference – three times as much axial isomer as expected – is much greater than is normal for semi-quantitative calculations of conformational equilibria of this sort in cyclohexanes (and corresponds to a difference in apparent A-value of 0.9 kcal mol^{-1}). And if we look at the corresponding equilibria (Table 1) for glucose derivatives with other electronegative substituents at the anomeric centre we see that the hydroxyl group is in fact the only one in the Table for which the equatorial isomer is preferred. Evidently an electronegative substituent at the anomeric position of glucose shows a distinct preference for the axial position, which generally outweighs the expected preference for the equatorial configuration.

This is the anomeric effect, and it is quite general. It is usually defined in terms of the preference of electronegative substituents at the anomeric centre of pyranoses for the axial configuration [7–10, 14–18]. A quantitative estimate of the magnitude of the effect must take into account the steric preference of a group larger than H for the equatorial position. This is assumed to be of the same magnitude as in a substituted cyclohexane (an assumption which neglects the differences between the carbocyclic and heterocyclic rings mentioned above [19], but does not lead to serious inconsistency in most cases). The

Table 1. Anomeric Equilibria of Glucopyranosides

Compound	*Substituent*	% *Axial Isomer*
D-Glucopyranose[a]	OH	36
Methyl-D-glucopyranoside[b]	OMe	67
Penta-O-acetyl-D-glucopyranoside[c]	OAc	86
Tetra-O-acetyl-D-glucopyranosyl chloride[d]	Cl	94

Conditions
[a] 25° in water [9a]
[b] 25° in methanol [9a]
[c] 25° in 1 : 1 acetic acid – acetic anhydride [11, 12]
[d] 30° in acetonitrile [13]

magnitude of the anomeric effect is thus defined [20] as the sum of the free energy difference for the equilibrium:

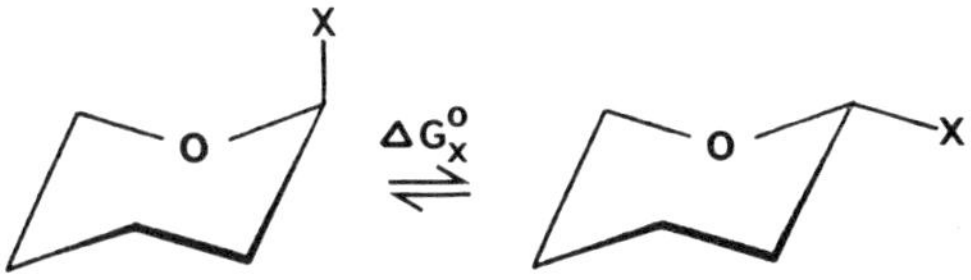

and the 'A-value' for the same substituent in cyclohexane:

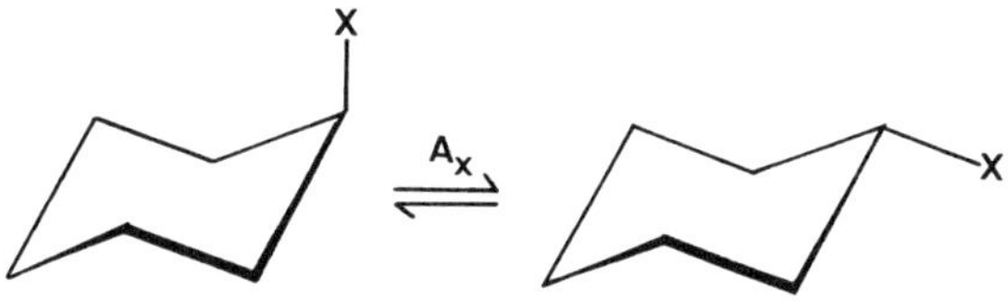

For example, the data given above for α and β-D-glucose allow an estimate of the anomeric effect of the hydroxyl group in water of 0.9 kcal mol^{-1} (using an A-value for OH of 1.25 kcal mol^{-1} [10a, 21]).

Table 2. The Anomeric Effect in Pyranose Sugars: Effects of Substituents on the Ring

Sugar		% Axial at Equilibrium	$\Delta G°$ (kcal)	Anomeric Effect (kcal)	Ref.
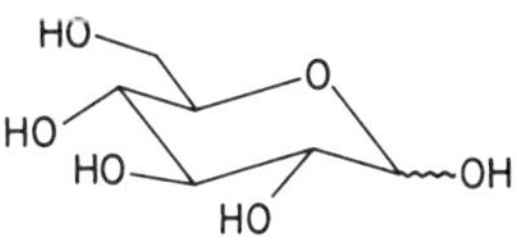 (D-glucose)		36	−0.35	0.9	[9, 10a]
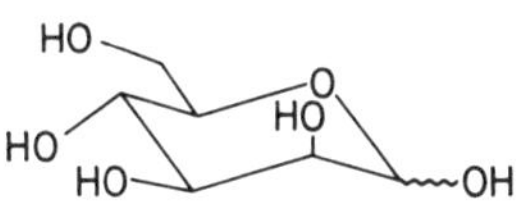 (D-mannose)		68	0.45	1.35	[9, 10a]
2-O-methyl-D-mannose		75	0.65	1.55	[22]
2,3-Di-O-methyl-D-mannose		80	0.86	1.76	[22]
2,3,4,6-Tetra-O-methyl-D-mannose		86	1.08	1.98	[22]
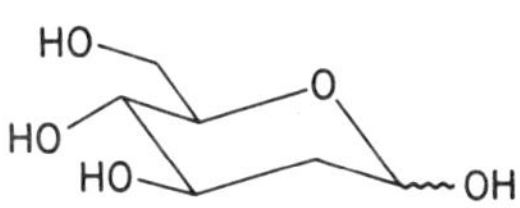		47.5	−0.05	1.20	[9, 10a]
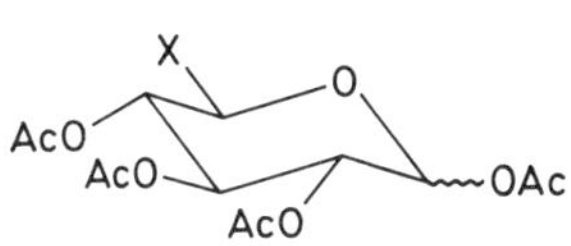	X = H			1.30	[12]
	CH_3			1.31	[12]
	CH_2I			1.35	[12]
	CH_2Cl			1.43	[12]
	CH_2OAc			1.45	[12]
	CH_2OTs			1.75	[12]

Given this figure one ought to be able to estimate anomeric and conformational equilibria for pyranose derivatives with reasonable accuracy, but in fact it is not possible to characterise the anomeric effect of a group by a single value which will serve in all cases. Similar calculations for α and β-D-mannose, which have axial 2-OH groups, give a value of 1.35 kcal mol^{-1}, and for the corresponding 2-deoxy-sugar the figure comes out as 1.2 kcal mol^{-1} [10a]. These results are summarised in Table 2, which shows also how the anomeric effect of the acetyl group depends on the electronegativity of the substituent at the 5-position in a series of substituted xylopyranose acetates [12]. Estimating the anomeric effect accurately enough to give reliable quantitative predictions of conformational equilibria in pyranose sugars is thus a complicated business, and the results are still not entirely satisfactory.

These difficulties can be largely avoided by working with less highly substituted systems. The simplest system which can exhibit the anomeric effect (as defined above in terms of axial preference) is a tetrahydropyran with an electronegative substituent at the 2-position, and most recent investigations have used such compounds. In fact the magnitude of the anomeric effect is not very different in the simpler systems and the sugar series, as shown by the comparative data in Table 3.

The data for the various methylated mannoses given in Table 2 show that the anomeric effect of the hydroxyl group in water is over twice as large in 2,3,4,6,-tetra-O-methyl-D-mannose as in D-glucose, thus suggesting that the effect may be sensitive to solvation. In fact, although there is a clear trend towards a lower anomeric effect in more polar solvents, the major difference is simply between the effect in organic solvents and in water. Table 4 shows data for the anomeric effect for 2-methoxytetrahydropyrans in a wide range of solvents, in order of increasing dielectric constant, which illustrate these points. (Note small systematic differences between 2-methoxytetrahydropyran and the 4 and 6-methyl derivatives in a given solvent [19].) It is clear that, with

Table 3. Axial Preferences in Simple Tetrahydropyran Derivatives Compared with those in Sugars [23]

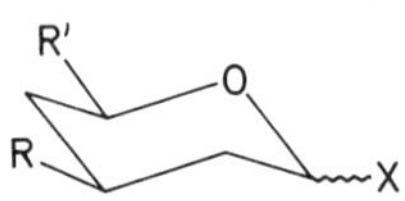

	% Axial isomer at equilibrium for		
	X = OH	OMe	OAc
R = CH_3, R' = H	34		
R = CH_2OH, R' = H	34		
R = CH_3, R' = H		69	75
R = H, R' = CH_3		67	75
R = H, R' = CH_2OH		70	75
D-xylosyl – X	36	69	79[a]
D-glucosyl – X	36	77	88[a]
D-galactosyl – X	36	71	79[a]

[a] These data [24] refer to fully acetylated sugars

Table 4. Solvent Dependence of the Anomeric Effect of the Methoxy Group in 2-Methoxytetrahydropyrans (MeOTHP)

Compound	Solvent	% Axial OMe	Anomeric Effect (kcal mol^{-1})[a]	Ref.
4-Me-2-MeOTHP	1,4-dioxan	77	1.31	[20]
6-Me-2-MeOTHP	1,4-dioxan		1.20	[36]
2-MeOTHP	CCl_4	82, 83	1.53	[17, 34]
6-Me-2-MeOTHP	CCl_4	77.3	1.31, 1.34	[19, 36]
4-Me-2-MeOTHP	CCl_4	80.2	1.41	[19]
2-MeOTHP	benzene	82	1.51	[17, 33]
2-MeOTHP	CS_2	80	1.41	[17]
2-MeOTHP	$CHCl_3$	71, 78	1.13–1.34	[17, 37]
6-Me-2-MeOTHP	tetrahydrofuran		1.20	[36]
2-MeOTHP	acetone	72	1.15	[17]
6-Me-2-MeOTHP	ethanol	65.5	0.96	[19]
4-Me-2-MeOTHP	ethanol	67.9	1.05	[19]
2-MeOTHP	methanol	69	1.08	[17]
6-Me-2-MeOTHP	methanol	62.5, 67	0.91–1.03	[17, 20]
4-Me-2-MeOTHP	methanol	67.3, 69	1.03–1.08	[19, 20]
6-Me-2-MeOTHP	nitrobenzene	67	1.03	[36]
2-MeOTHP	acetonitrile	68, 65	0.96–1.05	[17, 33]
6-Me-2-MeOTHP	acetonitrile	64.3	0.93	[19]
4-Me-2-MeOTHP	acetonitrile	67, 65	0.96–1.03	[19, 20]
2-MeOTHP	DMSO	74	1.22	[17]
2-MeOTHP	water	52	0.65	[17]

[a] Based on a constant A-value of 0.6 kcal mol^{-1} for the OMe group [35].

the result in dimethyl sulphoxide (DMSO) as the only significant exception, the anomeric effect is higher in less polar media and lowest in hydroxylic solvents. In this case only a part of the difference can be attributed to the increase in the apparent size of the OMe group due to tighter solvation in the H-bonding solvents, since its A-value increases only from 0.6 to 0.74 on going from CCl_4 to water [35]. (But see [36].)

Increasing the size of the alkoxy-group has remarkably little effect on the axial preference, as shown by the data in Table 5. In the absence of much information on the dependence of A-values on the size of the OR group, accurate anomeric effects cannot be calculated, but it is likely that they are not very different in the series methoxy, ethoxy, isopropoxy and *t*-butoxy. Making the alkoxy-group more electron-withdrawing, on the other hand, clearly increases the anomeric effect, from around 1 to about 2 kcal mol^{-1} on going from ethoxy to trifluoroethoxy. Conversely the effect is substantially reduced for the trimethylsilyloxy-group [36a].

This is clearly an expression of a general trend, which was already apparent in the data of Table 1, for strongly electron-withdrawing groups to have strong

Table 5. Anomeric Effect: Dependence on the Alkoxy Group in 2-Alkoxytetrahydropyrans

Group	% Axial in		
	Neat liquid[a]	CCl_4[b]	MeCN[b]
HO	77		
MeO	72	77–82	64–67
EtO	68	76–80	60–65
n-PrO		82	65
n-BuO		82	63
i-PrO	66	75	58, 60
t-BuO	62	67, 70	47, 52
$ClCH_2CH_2O$	77		
Cl_2CHCH_2O	88		
Cl_3CCH_2O	95		
F_3CCH_2O	92		
PhO	81		

Data from references [30][a], [19][b] and [34][b] for 2-alkoxytetrahydropyrans [30, 34] and their 4- and 6-methyl derivatives [19].

axial preferences, and thus large anomeric effects. Stoddart [9b] quotes the series

$$\text{halogen} > \text{PhCO}\cdot\text{O} > \text{AcO} > \text{AcS} > \text{AS} > \text{MeO} > \text{RS} > \text{HO} > \text{NH}_2 > \text{MeO}_2\text{C} > \text{pyridinium}$$

as a rough guide to the relative magnitudes of anomeric effects. The quantitative data available are given in Table 6, arranged in approximate increasing order of anomeric effect. The figures may be considered good to ±0.2 kcal mol^{-1} under the conditions given, in view of uncertainties associated with A-values and medium effects, and are accurate enough to show clearly that the biggest differences are those brought about by a change of element, with the largest anomeric effects associated with the halogens.

So large is the axial preference of any of the halogens, in fact, that it is the dominant factor in determining conformational equilibria in most pyranosyl halides. Where this preference is reinforced by the equatorial preferences of substituents at other positions one chair form is decisively favoured [37]. Where the two factors are opposed the anomeric effect wins for all but hexopyranose derivatives. Thus all the tri-O-acylpentopyranosyl halides exist preferentially (in solution) in conformations with the halogen atom axial [37–40] even where this means that all the acyloxy substituents are in axial positions also: as is the case for example for tri-O-benzoyl-β-D-xylopyranosyl fluoride [39], cloride [38] and bromide [41]. All these compounds have been shown to adopt the $^1C_4$ conformation (*13*) in solution almost exclusively, although the alternative $^4C_1$ form (*14*) would be expected to be preferred on steric grounds. The same is true for the the corresponding tri-O-acetyl fluoride and bromide, though in these cases the preference for the all-axial form is

Table 6. Magnitude of the Anomeric Effect in Pyranose and Tetrahydropyran (THP) Derivatives

Group	*Anomeric Effect*[a] (kcal mol^{-1})	*Conditions*	*Ref.*
$PO(OMe)_2$	0.56	6-$AcOCH_2$THP in benzene	[25]
OH	0.9–1.35[b]	Aldopyranoses in water	[10a, 26]
	1.0[c]	2-HO-4-MeTHP in CCl_4	[27]
	0.9[d]	2-HO-4-MeTHP in D_2O	[27]
OMe	1.2	Pentopyranoses in MeOH	[28]
	1.4	Hexopyranoses in MeOH	[29]
	1.3	2-MeO-4-MeTHP in dioxan	[27]
	0.9	2-MeO-4-MeTHP in 30 % MeOH – water	[27]
O-alkyl	1.4	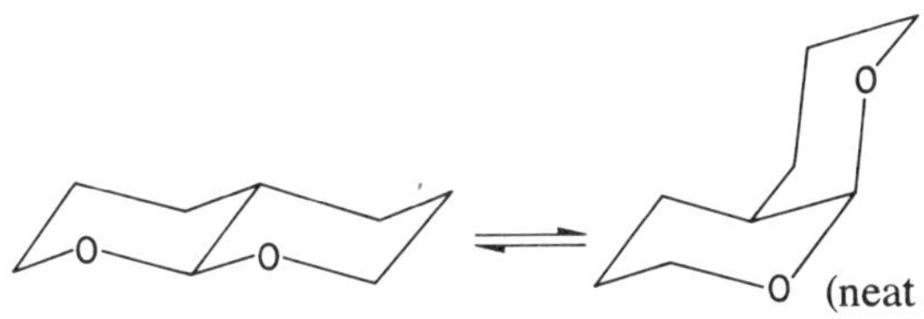 (neat liquid)	[29a]
OPh	~1.6[e]	2-PhO-THP (neat liquid)	[30]
OAc	1.3	Pentopyranose tetraacetates in 1 : 1-Ac_2O-AcOH	[12]
	1.5	Hexopyranose pentaacetates in 1 : 1-Ac_2O-AcOH	[12]
	1.35	2-AcO-4-MeTHP in AcOH	[27]
SMe	1.5[f]	2-MeS-THP in CCl_4	[31]
Cl	~2	Tetra-O-acetylglucopyranosyl chlorides in acetonitrile	[13]
	2.65[g]	2-Cl-4-MeTHP, neat liquid	[32]
Br	>3.2[h]	2-Br-4-MeTHP, neat liquid	[32]
I	>3.1[h]	2-I-4-MeTHP, neat liquid	[32]

[a] Defined as above (page 7)
[b] Details are given in Table 2
[c] Using an A-value of 1.0 for OH in CCl_4 [21]
[d] Using an A-value of 1.25 for OH in D_2O [10a, 21]
[e] Assuming an A-value of 0.7–0.8 for OPh
[f] Using Jensen's [33] A-value of 1.0 Kcal for SMe
[g] Lower in more polar solvents
[h] Limiting values. No equatorial anomer detectable in NMR.
The difference reflects only the different A-values for Br and Cl.

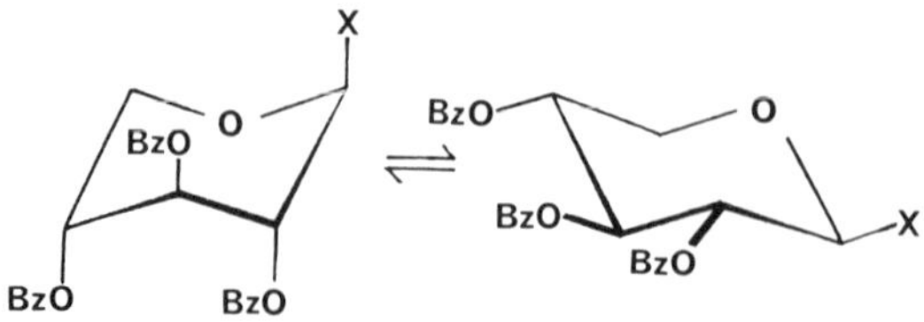

13 14

slightly reduced, and both compounds have been shown to exist in the all-equatorial form in the crystalline solid [42].

Recent results with other xylose derivatives suggest that the extraordinary dominance of the anomeric effect is somewhat exaggerated in these acylated systems, because the unfavourable non-bonded interactions between the 1,3-diaxial oxygen substituents are considerably smaller than expected. Significant amounts of all-axial isomer (*13*) are also present in solution for compounds where the anomeric effect must be smaller than that in the halides (X = PhCOO, MeO [43, 44]. These results are summarised in Table 7: the percentage of all-axial isomer (*13*) in solution clearly decreases with decreasing anomeric effect, but it remains significant even for compound (*13*, X = H) where the anomeric effect is absent [45]. This compound does at least take advantage of the absence of the anomeric effect to adopt the all-equatorial conformation in the crystal. None of the others do: in three cases the all-axial conformation is found also in the solid state. The two remaining cases (*13*, X = F, OMe) represent a compromise between anomeric and steric effects: both conformation (*13*), with its four axial substituents, and (*14*) with equatorial 2-X, are avoided in favour of the flexible (twist-boat) form (*15*) [42, 48]. It is perhaps surprising that this compromise conformation does not appear to occur in solution, but the NMR data on which the assignments are based are reasonably unambiguous.

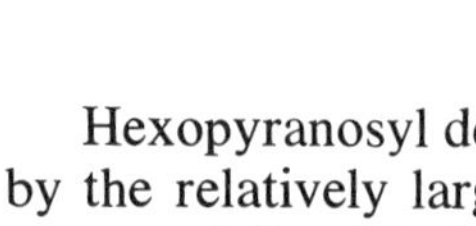

15

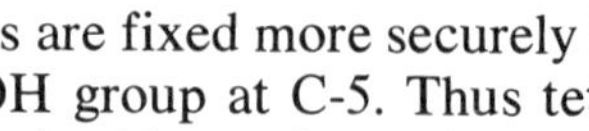

16

Hexopyranosyl derivatives are fixed more securely in the $^4C_1$ conformation by the relatively large CH_2OH group at C-5. Thus tetra-O-acetyl-β-D-glucopyranosyl fluoride (*16*) exists in this conformation, with all its groups equatorial (though there is a hint of conformational flexibility not observed for the α-fluoride, where steric and anomeric effects are in harmony) [49]. So too [50] does α-D-idopyranose pentaacetate (*17*), as is to be expected in view of the results with the pentopyranoses.

Table 7. Preferred Conformations of Tri-O-benzoyl-β-D-xylopyranosides (*13* ⇌ *14*)

Substituent, X	Conformation Observed		
	In Solution	In the Crystal	Ref.
F	~90 % (*13*)[a]	TB (*15*)	[41, 42]
Cl	~98.% (*13*)[a]	(*13*)	[41, 46]
Br	~98 % (*13*)[a]	(*13*)	[41]
OCOPh	~50 % (*13*)[b]	(*13*)	[43, 47]
OMe	~26 % (*13*)[b]	TB (*15*)	[44, 48]
H	~19 % (*13*)[b]	(*14*)	[45]

[a] in $CDCl_3$
[b] In d_6-acetone

The unexpectedly small steric requirements of O-acyl groups in these compounds, which result in favoured conformations such as (*13*) and (*17*), with two axial substituents on each side of the pyranose ring, even in relatively

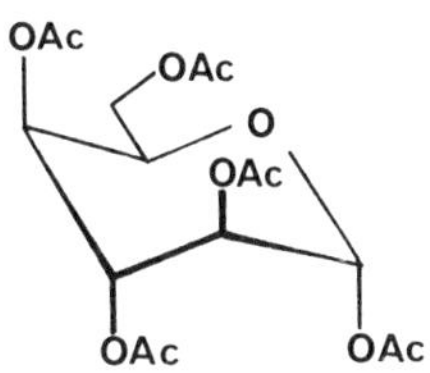

17

polar solvents like acetone and acetonitrile which generally reduce the anomeric effect, have been noted by many authors. Lemieux and Pavia [51] have suggested that the electrostatic repulsion between 1,3-diaxial oxygen substituents is reduced when they are acylated, because the electron density on oxygen is reduced. There is good evidence that 1,3-diaxial interactions between methoxy groups are substantially more severe: the tri-methoxy compound (*18*), for example, exists in the $^1C_4$ conformation shown in all solvents, whereas the diacetate (*19*) prefers the $^4C_1$ form, with all groups axial, also in all solvents (as do the corresponding dimesylate and di-p-nitrobenzoate in chloroform and dimethyl sulphoxide) [51].

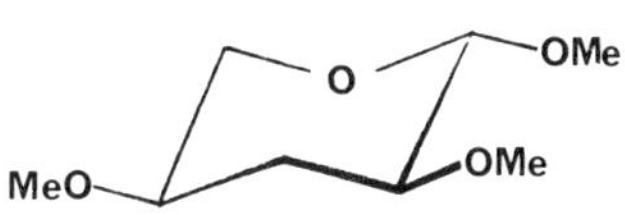

18

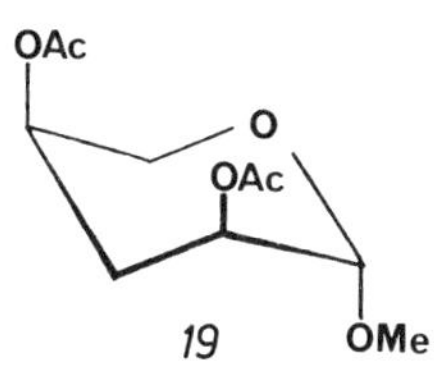

19

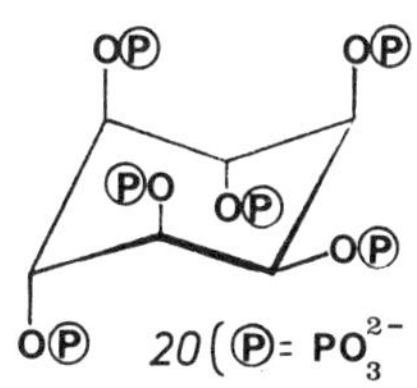

20 (Ⓟ = PO_3^{2-})

Suggestions that there are attractive interactions between 1,3-diaxial O-acyl groups are not supported by crystallographic evidence from all-axial benzoyl derivatives (the 1,3-diaxial interaction between two benzoyloxy groups is measurably larger than that between acetoxy groups) [42]. Nor do they seem necessary, in view of the recent observation that the dodeca-anion of myoinositol hexaphosphate in water prefers the conformation (*20*, $P = PO_3^{2-}$) with *five* phosphate dianion groups axial [52], three of which are on the same side of the ring (The hexaanion, which might be expected to have attractive (H-bonding) interactions between phosphate groups exists as the other isomer, with five monohydrogen phosphate groups equatorial.)

For elements from the first row of the periodic table the anomeric effect in tetrahydropyran and pyranose systems decreases with decreasing electronegativity, and is clearly greatest for $F > O > N > C$. For the most electronegative type of carbon, the sp-hybridised C of an acetylene, a small axial preference is apparent for 2-substituted 1,3-dioxanes (*21*) in non-polar solvents, though this is reversed in acetonitrile [53].

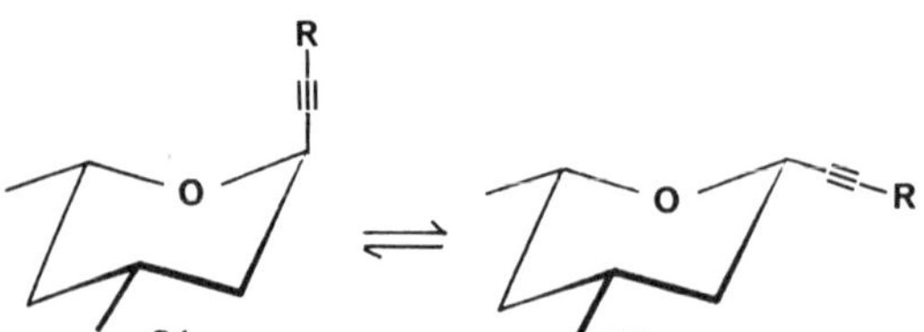

21a ⇌ *21e*

In the case of N the effect depends on the substituents at the nitrogen centre. Quantitative measurements of the anomeric effect of the sort discussed above (Table 6) are not available, because A-values have not been measured for the wide range of groups involved, but measurements by NMR of the position of conformational equilibrium for N-substituted N-pentopyranosides [54] have established an order of axial preference as follows:

$$\overset{-}{N}-\overset{+}{P}Ph_3 > (OAc) > \overset{-}{N}-\overset{+}{N}{\equiv}N > NHCOCF_3 > NHCOAr > NHPO(OMe)_2 \simeq NH_2 \simeq NHAc \simeq NHPPh_3 > \text{imidazolium} > \text{pyridinium}.$$

Here too the effect is largest for the more electronegative sp and sp^2-hybridised nitrogen groups, which are also those with the greatest charge density on N; and comparison with the conformational equilibria of the pentopyranosyl tetraacetates shows that the axial preference of the phosphinimine group $\overset{-}{N}\overset{+}{P}Ph_3$ is actually greater than that of acetoxy. Thus in the tri-O-acetyl-α-D-lyxose derivatives (*22*) the compound with $X = \overset{-}{N}\overset{+}{P}Ph_3$ exists in $CDCl_3$ as an equilibrium mixture containing 83% of the isomer with X axial, compared with 71% for the tetraacetate (X = OAc) and 52% for the azide ($X = N_3$) [53].

On the other hand, the corresponding compounds with $X = NH_2$, NHAc and $NHPPh_3$ exist almost exclusively in the $^1C_4$ form (*22e*) et equilibrium.

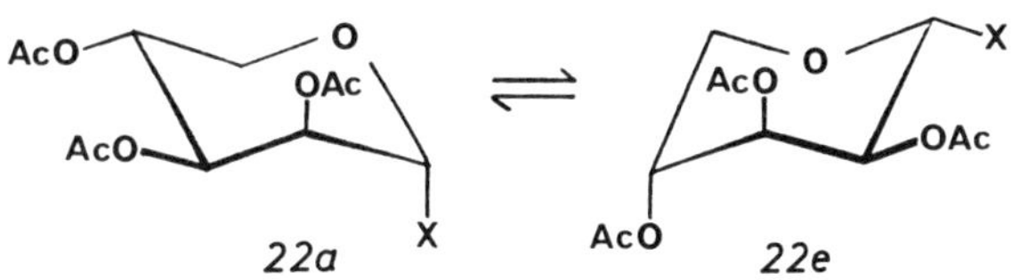

22a ⇌ 22e

This, and other, evidence shows that the anomeric effect disappears as one goes down the above series, so that groups such as NH_2 and NHAc show the usual steric preference for the equatorial configuration. At the end of the series is a group of substituents based on positively charged nitrogen which appear to show even smaller axial preferences, that is to say an enhanced preference for the *equatorial* position. This effect is the subject of the following Section.

3. The Reverse Anomeric Effect

Measurements of anomeric and conformational equilibria have established that the anomeric effect decreases in the (approximate) order:
Halogen > PhCOO > AcO > AcS > MeO > RS > HO > NH_2 > MeO_2C > pyridinium (page 10).
The substituents at the head of this series show a powerful preference for the axial position in 2-substituted tetrahydropyrans: those at the end of the list do not. Groups such as N-pyridinium appear to show a preference for the equatorial position, and there is evidence to suggest that this preference is stronger than the usual steric preferences found in cyclohexanes. This is the reverse anomeric effect.

The best evidence comes from systems where the 2-substituent is the N^+ atom of a heterocycle. Lemieux and Morgan [55] prepared several α-D-glucopyranosyl pyridinium systems (e.g. *23*), which might be expected to have axial pyridinium substituents:

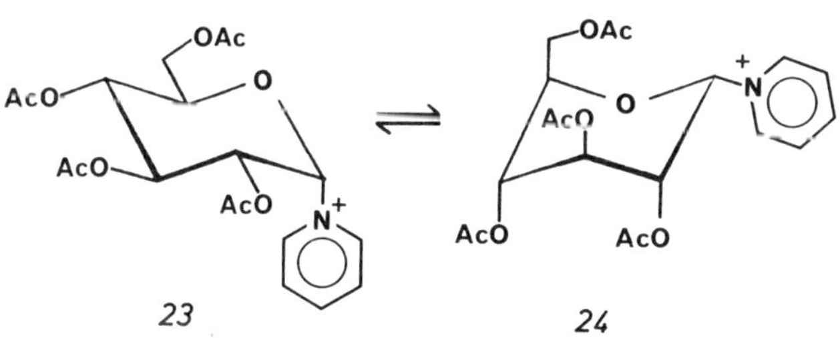

23 ⇌ 24

The (first order) ^1H NMR spectra, however, were clearly incompatible with structure (*23*), and the coupling constants of the protons on C(1) and C(2) in particular (J_{12} 2.8 Hz) suggested rather that these compounds existed in the $^1C_4$ conformation (*24*). A crystal structure determination [56] (details not yet published) of the 4-methylpyridinium compound showed a more complicated conformation (*25*) close to a boat. This allows the substituents on C(4) and

C(5) to remain (pseudo)equatorial (the barrier to complete ring inversion of the glucopyranose system is so large that even β-glucosyl fluoride retains the $^4C_1$ conformation, with F equatorial), while the acetoxy-groups on C(2) and C(3) are forced into (pseudo)axial positions.

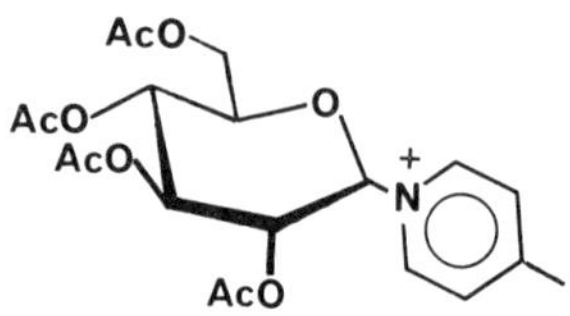

25

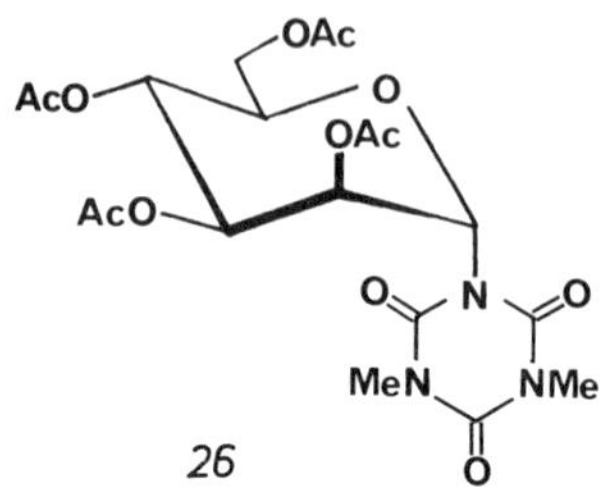

26

More recent results [57] on the related 1-(α-D-mannopyranosyl) isocyanuric acids show a situation similar enough to shed more light on the behaviour of (*25*). Compound (*26*) exists in the crystal in a modified $^4C_1$ conformation, with the C(1) end flattened but the aglycone still recognisably axial. But in solution the NMR parameters are consistent only with a flexible conformation (*27*).

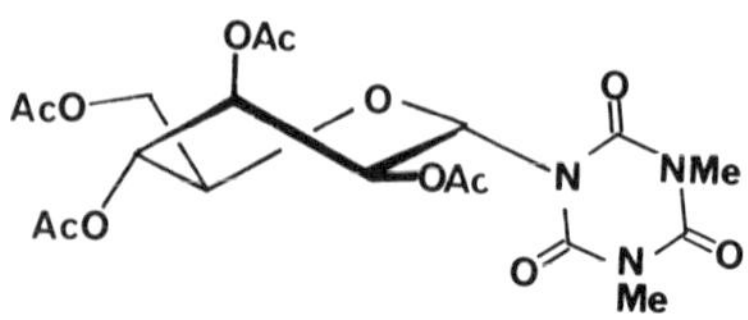

27

The coupling constants between the protons at C(3), C(4) and C(5) are very similar for (23 ⇌ 24 ⇌ 25) and for (26 ⇌ 27) suggesting that both compounds exist in solution in flexible conformations.

Both this result and the crystal structures show that conformations with heterocyclic nitrogen in an axial position are strongly disfavoured, particularly for the system where nitrogen carries a full positive charge.

These aglycone groups are very large, and the simplest explanation would be that their normal steric preference for the equatorial position outweighs the small anomeric effect expected for nitrogen substituents. Particularly, as we saw above, since 1,3-diaxial interactions between acetoxy groups do not appear to be as severe as originally thought. However, this explanation cannot account for the observations of Lemieux [58] and Paulsen et al [54], on the

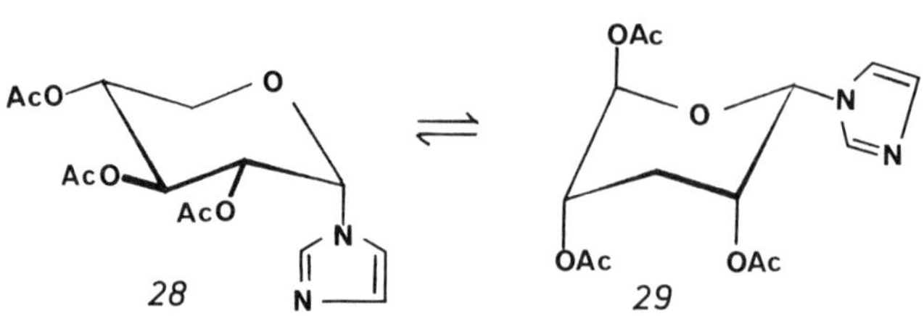

28 *29*

conformational behaviour of N-imidazolium glycosides. The more recent and more extensive evidence concerns the α-D-xylo-compound (*28*).
The equilibrium mixture in CCl_4 contains 35 % of the $^1C_4$ form (*29*), with the imidazole group equatorial and the three acetoxy groups axial, evidence that imidazole nitrogen also shows a marked preference for the equatorial position. This preference is greater in more polar solvents (65 % of (*29*) in $CDCl_3$, 85 % in d_6-acetone). But when sufficient trifluoroacetic acid is added to the $CDCl_3$ solvent to protonate (*28*$\rightleftharpoons$*29*) essentially completely, the NMR spectrum shows that *only* the $^1C_4$ isomer (*30*), with the imidazolium group equatorial, is present (as is also the case with the N-pyridinium compound) [54].

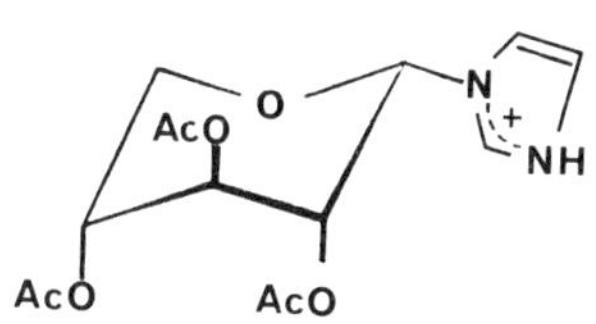

30

Clearly the steric requirements of the heterocyclic ring have not changed on going from (*29*) to (*30*). The increased equatorial preference of the protonated imidazole must therefore be an electronic effect, resulting from the partial positive charge on nitrogen in (*30*).

The reverse anomeric effect thus appears to be a real phenomenon, though because it occurs in a very limited range of compounds its main importance is as a test of theories of the anomeric effect. To date none of these explain the reverse anomeric effect very well (see below, page 71). It is however possible to make a useful generalisation about the relative stabilities of conformations of N-glucopyranosides: apart from a few special cases, such as $\overset{-}{N}\overset{+}{P}Ph_3$ and N_3, which show normal anomeric effects, neutral nitrogen substituents behave much as they do in cyclohexanes, with large groups showing a strong preference for equatorial positions; but when the nitrogen atom bears a positive charge this steric preference is reinforced by the reverse anomeric effect. (Apart from the results cited specifically in this Section, the conformation of a number of related N-glycopyranosides are consistent with this generalisation [54, 59–62]).

The identification of this effect of a positive or partially positively charged nitrogen substituent has encouraged a search for similar effects in compounds with a carbon atom bearing a partial positive charge at the 2-position of tetrahydropyrans [53, 63, 64]. The groups considered have been C≡N (see above, page 14) CO_2Me, CH_2Cl, CH_2Br and substituted phenyl; but the results are not clear-cut, and show mainly that these groups do not exert a significant anomeric effect, reverse or otherwise.

4. The Exo-Anomeric Effect

The anomeric effect is usually defined as the preference for the axial configuration of electronegative substituents at the 2-positions of tetrahydropyran rings. In terms of the conformation about the ring-O-C(2) bond this is equivalent to a preference for the gauche (*synclinal, 31*) over the *antiperiplanar* conformation (*32*).

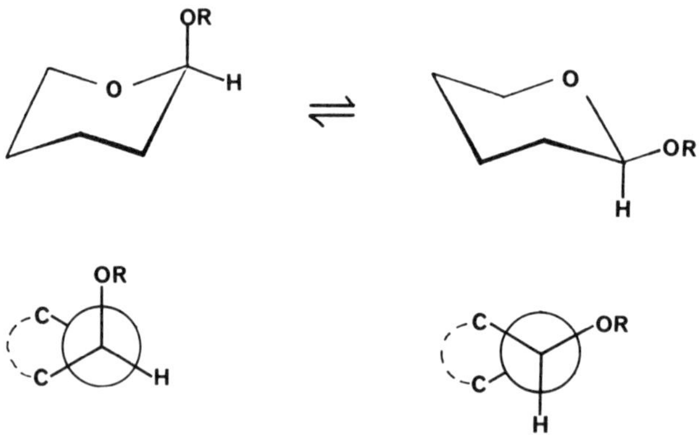

31 *32*

The exo-anomeric effect refers to the related preference for the gauche conformation about the O−R bond of the substituent group (the aglycone in the case of sugars) [65].

Three staggered conformations are possible for rotation about this bond in both the axial and equatorial forms of a 2-alkoxytetrahydropyran. These are referred to as A1–3 and E1–3, respectively.

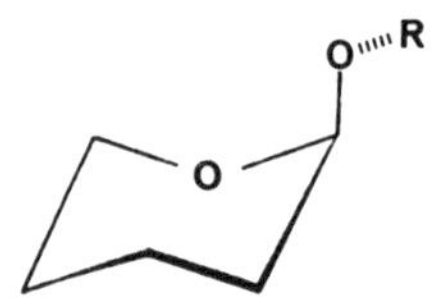

A1

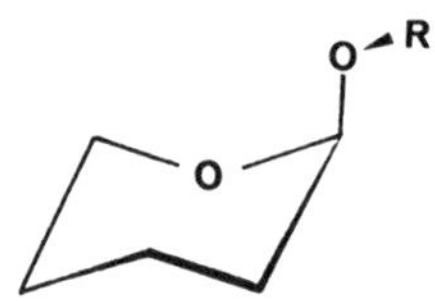

A2

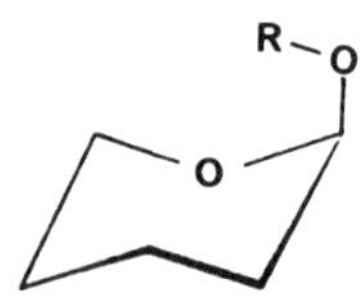

A3

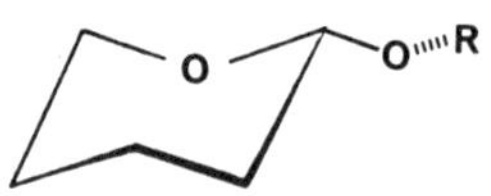

E1

E2

E3

In the light of what we have learned about the anomeric effect the conformational analysis of the two systems is straightforward. In each series one conformer has a more or less severe steric interaction: in isomer A3 the R group lies above the ring in close proximity to the two axial hydrogen atoms; and in conformer E2 it is eclipsed, and thus suffers a 1,3-diaxial type repulsion with the axial H on C(3). In both cases, therefore, the choice is likely to lie between a gauche (A1, E1) and an antiperiplanar (A2, E3) conformation. Given the clear preference for gauche conformations arising from the anomeric effect, A1 and E1 are expected to be the preferred conformations of axial and equatorial 2-alkoyltetrahydropyrans and alkyl glycosides.

Experimental evidence initially proved difficult to obtain, but has gradually accumulated over the years to the point where it is reasonable to accept that this expectation is generally correct. The early attempts involved comparing the observed and calculated dipole moments for simple 2-alkoxytetrahydropyrans, using the amounts of axial and equatorial forms estimated from the coupling constants of the anomeric proton [17, 34]. These comparisons were consistent with the A1 conformation for the axial isomers, and either E1 or E2 (which would have identical dipole moments) for the equatorial isomers. Lemieux and his co-workers could find no evidence for a significant amount of the E2 conformer in several systems [66], and this result seems to be general [67].

The simplest NMR method directly applicable to the problem is the measurement of the coupling constant between the anomeric proton and the α-carbon atom of the (aglycone) R group. Vicinal $^{13}C-^{1}H$ coupling constants do show a dependence on torsion angle similar to the familiar Karplus equation [67, 68], but direct assignments of torsion angles from the coupling constants cannot yet be made without important assumptions (for example, that torsion angles in solution are equal to those in the crystal). Nevertheless, the conclusions from this work, from the constancy of the chemical shift of the anomeric carbon, and from some nOe measurements, generally support the original conclusion that A1 and E1 represent the favoured conformations of the axial and equatorial forms.

For molecules in the solid state the evidence is unequivocal. It was noted early on that alkyl pyranosides adopt the A1 or the E1 conformation, as appropriate [34] and a recent analysis of over 50 carbohydrate structures [69] reveals the following regularities: for axial methyl pyranosides the torsion angle $O-C-O-CH_3$ (nominally 60° in conformer A1) lies between 61 and 74°. For the equatorial anomers the range is 68–87°. Oligosaccharides show a wider range of variation (75–121° and 71–105° for axial and equatorial isomers, respectively, no doubt because larger crystal-packing forces can be brought to bear on larger aglycones: but the range is much smaller than that, for example, for rotation about the adjacent (O−R) bond [69]. Thus all the evidence is consistent with a significant restriction of rotation about the anomeric-C−OR bond, which is of prime importance in determining the conformations of glycopyranosides and 2-alkoxytetrahydropyrans [65, 67, 70, 71].

There is conflicting evidence as to whether the exo-anomeric effect is larger for axial or equatorial OR groups: even the analysis of crystal structures quoted above does not give a clear answer for glycopyranosides in the solid state, and results in solution are similarly ambiguous, particularly for complex oligosaccharide systems [72–75]. What is clear is that it is of comparable importance in the two cases, and is the dominant short-range interaction controlling the conformation about the glycosidic linkage in both α and β-linked oligosaccharides; though in long-chain and branched systems the build-up of longer-range interactions may diminish the preference for A1 or E1-like conformations [73].

A particularly clear illustration of the working of the exo-anomeric effect comes from the work of Deslongchamps and his co-workers on the spiro-acetal (*33*) [76]. This compound has in principle three distinct conformational isomers based on all-chair tetrahydropyrans, namely (*33a, b* and *c*), corresponding to conformations A1, E1 plus A2 and E3, respectively, of the 'O–R' bond of the second ring.

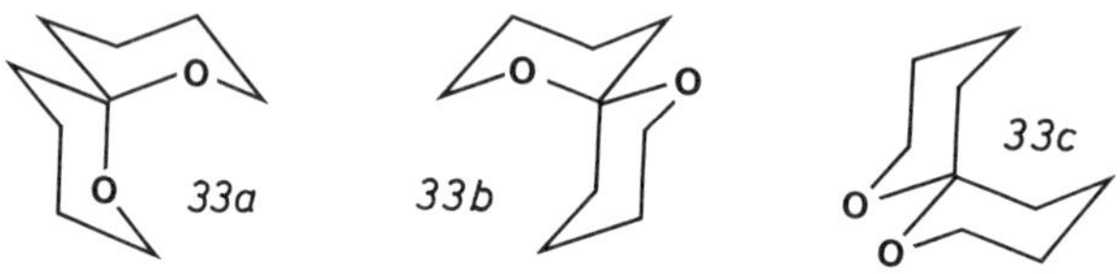

33a — A1   33b — E1, A2   33c — E3

The ¹³C NMR spectrum of (*33*) [76, 77] only shows five signals, ruling out (*33b*), and is best interpreted in terms of the A1 form (*33a*). (The assignment is consistent with X-ray structure determinations of two substituted derivatives of *33* [78, 79]). There is no indication of significant amounts of any other conformer, showing that the stereoelectronic effects at the acetal centre must be substantial. Here the anomeric effect, the exo-anomeric effect and the preference of the tetrahydropyran rings for the chair conformation all act in the same direction to produce a system which appears to be effectively conformationally rigid [76].

5. The Generalised Anomeric Effect

Although the anomeric effect, by definition, applies to the properties of cyclic compounds, our discussion of the exo-anomeric effect has made it clear that the factors responsible operate equally effectively on the conformation of the exocyclic O–R bond of a tetrahydropyran acetal or glycoside. It is to be expected, therefore, that similar effects will be observed in acyclic compounds. Thus the simplest acetal, dimethoxymethane, has been shown to exist in the gauche, gauche (+*sc*, +*sc*) conformation (*34*) [80, 81] as have chloral hydrate (*35*) [82] and polyoxymethylene $(CH_2O)_x$ [83]. Even propargyl alcohol and

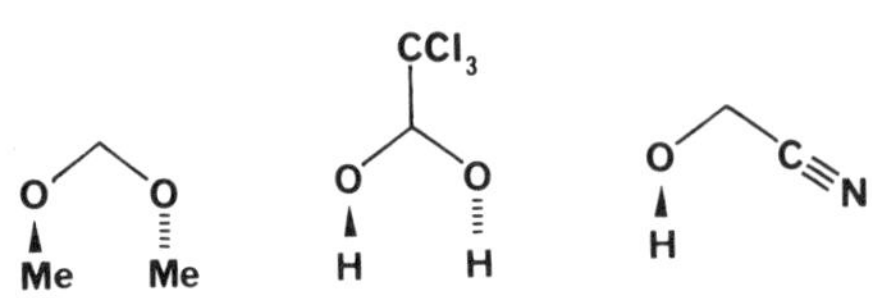

34   35   36

hydroxyacrylonitrile (*36*) favour the gauche conformation about the C–O bond in the gas phase [84].

The general preference for the gauche conformation about a carbon-heteroatom bond in systems R–X–C–Y has been termed the generalised anomeric effect by Lemieux [58] and Eliel [53, 85]. The original definition by Lemieux is very broad, and refers to "phenomena which are the result of the same kind of interactions as were proposed for explaining the anomeric effect, but present in non-carbohydrate structures" [58]. A broad definition is appropriate, because stereoelectronic effects related to the anomeric effect, in which two electronegative atoms interact through a central carbon *or other atom* are of importance in a wide range of organic and inorganic structures. We will discuss a number of systems where an axial preference may be evidence for the working of the generalised anomeric effect, starting with systems close to tetrahydropyran.

A great many cases are known where substituents on six-membered rings prefer an axial position [86], and not all of these are related to the anomeric effect. A case in point is the 2-halocyclohexanone system [87], where the axial preference clearly decreases in the order Br > Cl > F, and can be explained as a combination of steric effects and dipole-dipole interactions. So the moderate preference of the 2-alkoxy-group for the axial conformation in 2-alkoxycyclohexanones (*37*), which is comparable in magnitude to that caused by the anomeric effect in 2-alkoxytetrahydropyrans (Table 8) is not surprising. But when the alkoxy-group is adjacent to both a carbonyl group and a ring oxygen atom the axial preference appears to be substantially enhanced (compound *38*) [88], apparently by a stereoelectronic effect.

When the exocyclic double bond is that of a simple olefin (*39*) rather than carbonyl, the preference of a 2-alkoxy-group for the axial conformation is clearly greater, even though the important dipole-dipole interaction has been removed [89]. That this is a stereoelectronic effect is shown clearly by the result with the enol ether (*40*): evidently the π-system is a net donor in the interaction. Similar results have been found for alkoxydihydropyrans (*41 vs. 42*), where the double bond is endocyclic [89].

These are of course allylic systems, but their conformational behaviour is similar enough to that of 2-alkoxytetrahydropyrans to suggest that the generalized anomeric effect is operating.

When both heteroatoms of the R–X–C–Y system are in a six-membered ring the generalised anomeric effect is expressed differently. 1,3-Dioxans have a strong but otherwise unremarkable preference for the chair conformation [5a]. They are more puckered than cyclohexanes and an axial substituent in the 2-position (*43*) therefore suffers more severe 1,3-diaxial interactions than

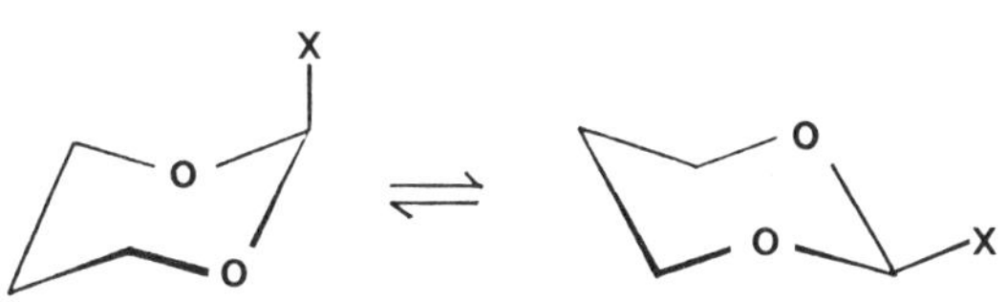

43

Table 8. Axial Preferences for methoxyl groups adjacent to sp^2-hybridised ring carbon

		% axial in CCl_4
37		63
		69
38		100
39		78
40		94
41		a
42		b

[a] Not determined quantitatively, but the pseudoaxial isomer clearly predominates.
[b] Not determined quantitatively, but more pseudoaxial isomer than for [41].

in cyclohexane. Nevertheless the 2-methoxy-compound (*43*, X = OMe) shows a clear preference for the axial conformation, as would be expected for a system experiencing a double anomeric effect.

In rings containing at least one nitrogen atom the generalised anomeric effect controls the axial-equatorial equilibrium of the exocyclic group attached to N. In 1-oxa-3-azacyclohexanes (*44*), for example, the predominant

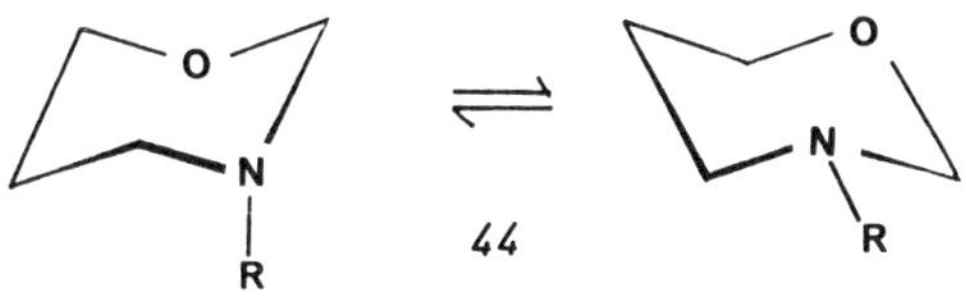

44

conformer has the N-alkyl group axial [5a, 90, 91]. One N-alkyl group is also axial in the favoured conformations of 1,3-diaza-cyclohexanes (*45*) [5a, 92, 93] and 1,3,5-triazacyclohexanes (*46*) [94], even when R = *tert.*-butyl [95].

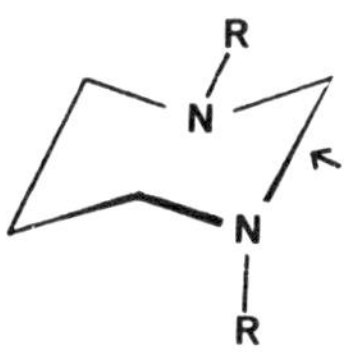

45

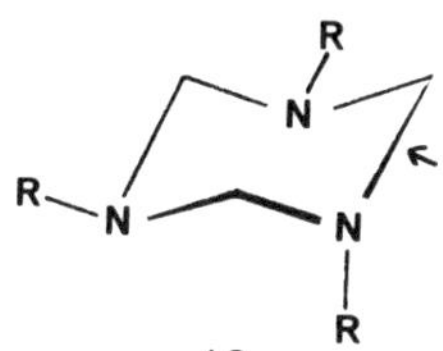

46

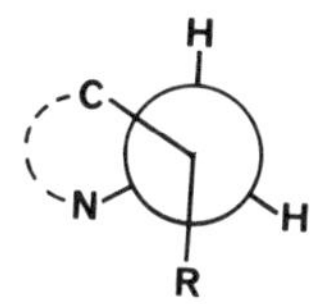

47

In each case the axial N-alkyl group is gauche to the ring C−N (or C−O) bond (*47*), as expected if the conformation about the bond indicated (*45, 46*) is controlled by the generalised anomeric effect. (Conformations of (*45*) and (*46*) with more than one axial R group, though favoured by this stereoelectronic effect, would suffer severe 1,3-diaxial repulsions.) No doubt electronegative substituents at the 2-positions of N-alkylpiperidines would also favour the axial conformation, but such compounds are generally too reactive to be prepared. The most closely relevant example is the nitrile (*48*), one of two related examples based on the 1-oxa-2-aza-cyclohexane system shown to prefer the conformation with the CN group axial [95a].

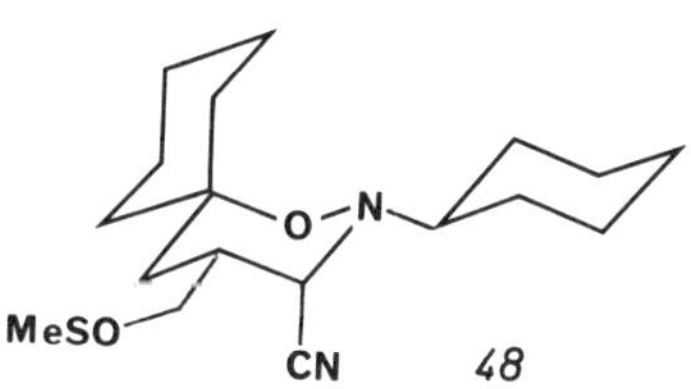

48

a. Sulphur Compounds

The conformational properties of six-membered rings containing sulphur have been studied much less than tetrahydropyran systems [5b], but it is clear that a strong generalised anomeric effect operates in systems (*49*) with an electronegative substituent at position 2. Axial preferences are magnified, relative to the tetrahydropyran system, because 1,3-diaxial repulsions are smaller as a result of the long C−S bonds, and this factor is especially important in 1,3-dithanes (*50*) and 1,3,5-trithianes (*51*).

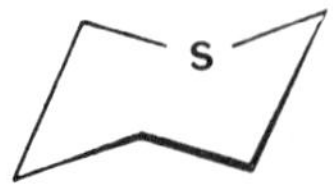

49

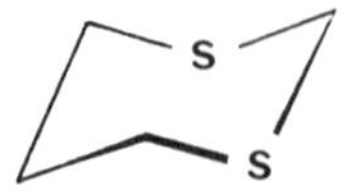

50

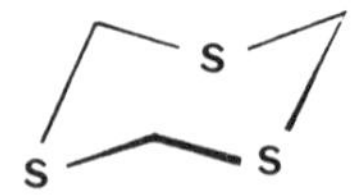

51

The first evidence was the demonstration that *trans*-2,3-dichloro-1,4-dithianes (*53*) and thioxanes (*52*), like the corresponding dioxanes, exist in the crystal and in solution [96, 97] in the diaxial conformation.

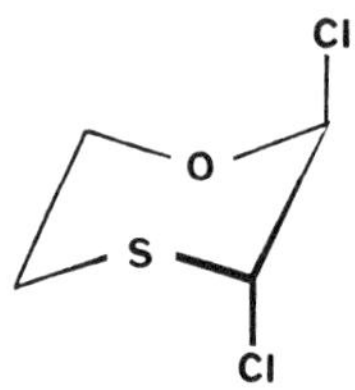

52

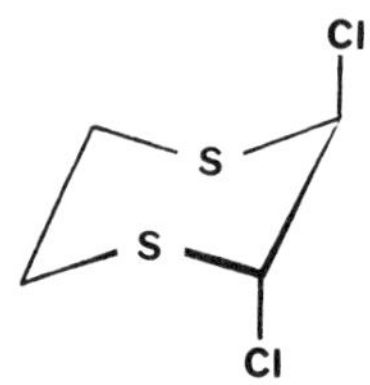

53

2-Alkylthiotetrahydropyrans were also shown to exhibit the exo-anomeric effect, and to exist preferentially in the expected (A1, E1 or E2) conformations in the axial and equatorial forms [98].

The best quantitative data come from ^{1}H NMR measurements of the axial-equatorial equilibrium for a series of 2-substituted thianes [99]. The results are summarized in Table 9, and show that the axial preferences of alkoxy-groups are distinctly larger than found in 2-alkoxytetrahydropyrans, though those of 2-Cl and 2-alkylthio-groups are smaller. (Anomeric effects cannot be calculated accurately from these data, because of the absence of satisfactory A-values for the sulphur-containing ring, though scattered data

Table 9. Axial Preferences of 2-substituents in Thiane Rings [99] in CCl_4

Compound		$\Delta G°_{e-a}$ (kcal mol^{-1})
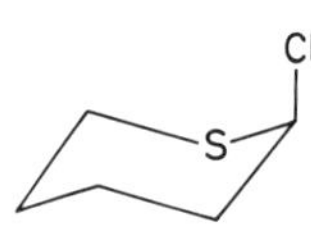		1.75
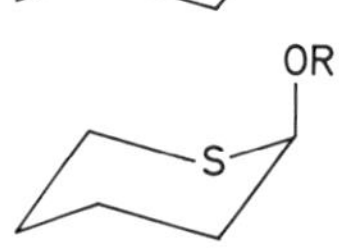	R = Me	1.53
	R = *i*-Pr	1.35
	R = *t*-Bu	0.74
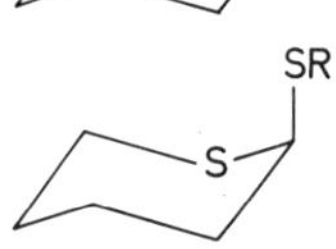	R = Et	0.42
	R = *i*-Pr	0.25
	R = *t*-Bu	−0.19

leave little doubt that alkyl substituents retain the expected equatorial preference.) Results from other laboratories are generally consistent with the data in Table 9, and include measurements on 2-alkoxythianes (around 80% axial) [100] and on several 2-alkylthiothianes and 2-phenylthiothiane [101, 102], which show comparable amounts of axial and equatorial isomers. Axial preferences are larger in dithiane (*50*) and trithiane (*51*) systems. Although 2-phenylthiothiane exists as a close to 50 : 50 mixture of axial and equatorial forms [101, 102], 2-phenylthio-1,3,5-trithiane is nearly 90% axial (*54*), and even

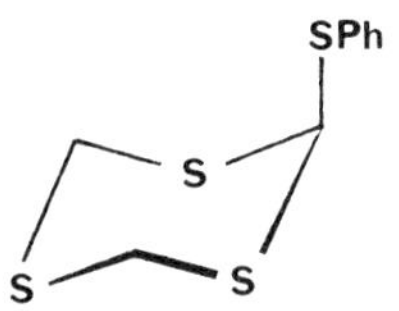

54

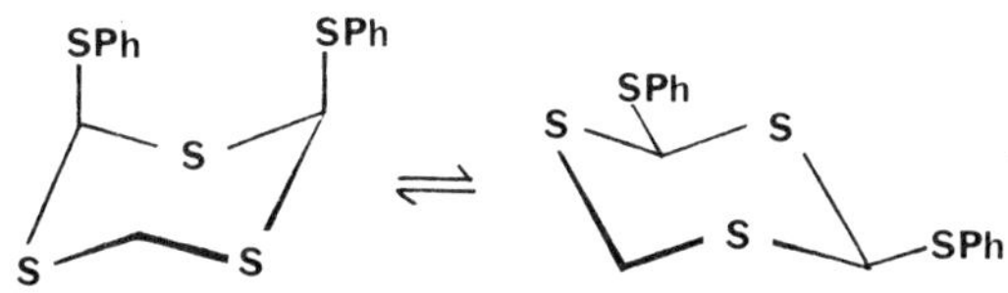

55

the 2,4-diphenylthio-compound (*55*), though predominantly diequatorial, shows nearly 25% of the diaxial isomer [103]: while the corresponding benzoyloxy compounds are essentially all axial, or diaxial, respectively [103]. It is interesting that in these sulphur-containing rings the carboxyl group also shows a very clear axial preference, in contrast to the complex picture found for the tetrahydropyran-2-carboxylic acid system (see above, page 00). The effect is larger for the esters, and larger also for 1,3-dithianes than for the trithianes, as illustrated by the data below (NMR measurements at −80° in

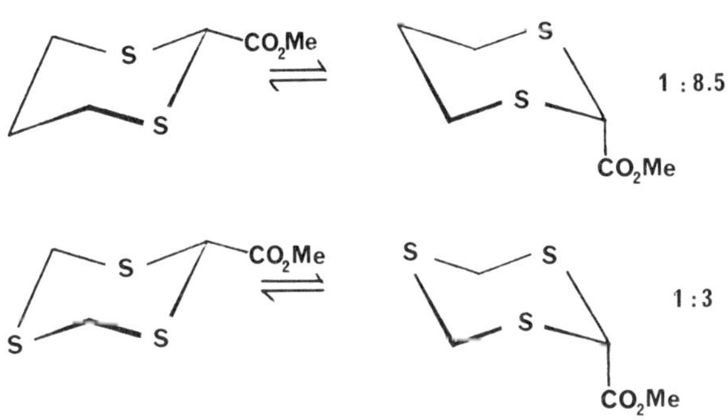

d_6-acetone) [104]: and considerably larger also in less polar solvents like CCl_4. Even the 2-phenylethynyl group shows an axial preference in these systems [104].

The most recent results on 2-substituted thianes concern the properties of the spiro-compounds (*56*, X = O and X = S) [105].
These compounds could exist in three or four (b_1 and b_2 are equivalent for *56*, X = S), all-chair conformations, but only isomer (*56a*) is actually observed [105]. This is the result expected if the anomeric and exo-anomeric (or second anomeric) effects are similar in magnitude (about 1.4 kcal mol^{-1}) in (*56*) and

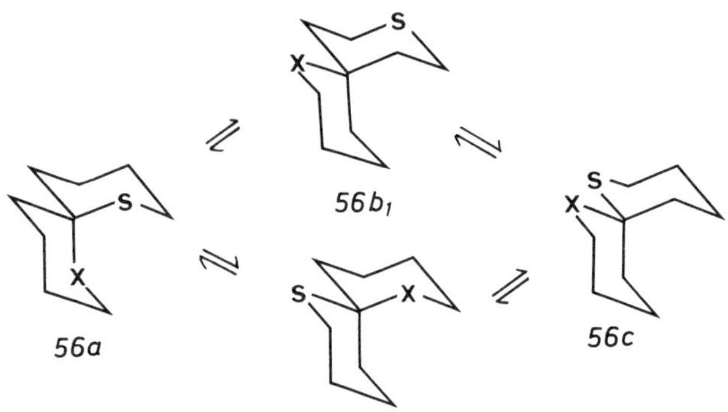

56a 56b₁ 56b₂ 56c

the corresponding acetal (*33*) (page 20, above). Detailed results on several substituted derivatives of (*56*, X = S and X = O) suggest that the anomeric effect is slightly greater in the monothioacetal (*56*, X = O) than in the dithio-compound (*56*, X = S), consistent with previous data summarised in Table 9 (above).

b. Transmission through Phosphorus

The best evidence that the generalised anomeric effect can be transmitted through centres other than carbon comes from extensive studies on the conformations of six-membered rings containing phosphorus and sulphur. In some cases the preference of substituents for axial conformations clearly exceeds that at the corresponding carbon centres, but interpretation of the data requires some care because the usual steric preference of substituents for equatorial positions is reduced when they are attached to phosphorus. Thus the series of phosphines (*57*, R = Me, Et, *i*-Pr, *t*-Bu and Ph) shows no strong conformational preference, except for (*57*, R = *t*-Bu), which exists solely in the equatorial conformation (*57e*) [106].

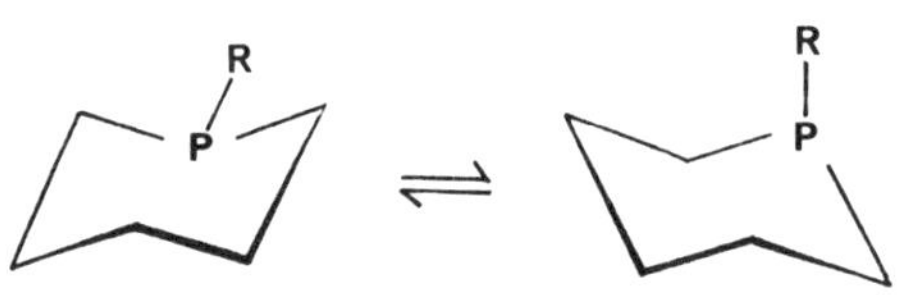

57e 57a

For smaller R-groups both isomers are present, with the equatorial predominant only at low temperatures [106–108]. At room temperature and above the axial form (*57a*) is preferred. Cyclic esters of trivalent phosphorus acids, on the other hand, such as (*58*) and (*59*) show a clear preference for the axial conformation when X is an electronegative group (F, Br, Cl, OMe, OBu,

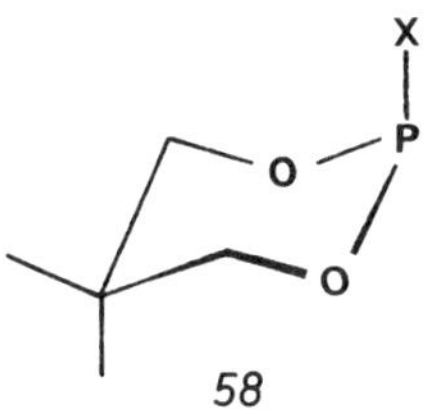

58

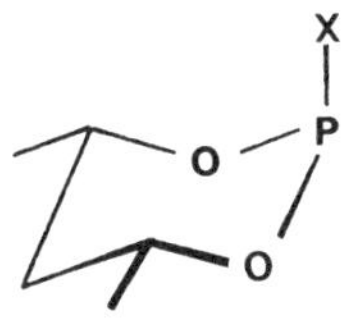

59

SPh) [109], as might be expected if the generalised anomeric effect were operating. But small alkyl substituents also prefer the axial position, as shown by the data in Table 10.

Table 10. Free-energy differences for the equilibrium

Group	$\Delta G°$ (kcal mol^{-1})[a]
Cl	1.5
MeO[b]	1.4
Me	0.99
i-Pr	0.63
t-Bu	−1.5
Ph	1.3
Me_2N	−0.94
MeNH	−0.12

[a] By NMR in d_6-benzene at ambient probe temperatures [110]
[b] In $CDCl_3$

The axial preference of Cl and OMe is greater than at the 2-position of a 1,3-dioxan, but so too is that of the methyl and other alkyl groups [110]. This is because another stereoelectronic factor affects the conformation about the P−O bonds: vicinal lone-pair-lone-pair interactions also favour the gauche conformation (see below, page 32), leading to an axial preference for the substituent on phosphorus. This factor can clearly be overcome by sufficiently large steric effects, as in the case of the *t*-butyl group. It is also overcome by what may be a (generalised) reverse anomeric effect in the case of the dimethylamino-group, which shows a clear equatorial preference [110, 111] of the sort already noted in our discussions of the reverse anomeric effect (see above, page 15). (That the effect is not simply steric is shown by the contrasting behaviour of NMe_2 and isopropyl derivatives (Table 10), and by the axial preference of PhNH in a similar system [112], where MeNH still prefers to be equatorial.)

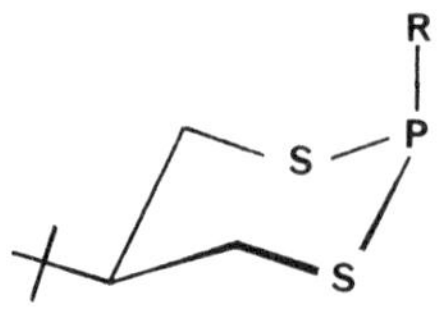

60

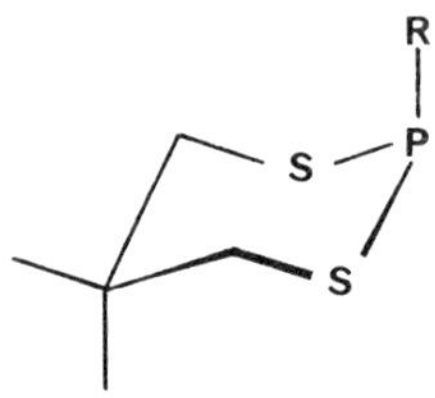

61

There is evidence for similar conformational preferences in cyclic esters (*60, 61*) of thiophosphorous acid [113].

The conformations of cyclic esters of trivalent phosphorus acids thus appear to be controlled by four main factors: the usual steric and dipole-dipole interactions, the gauche effect which operates on conformations about single bonds between two hetero-atoms both bearing lone-pair electrons [114], and the anomeric effect. The first of the two stereoelectronic effects is absent in cyclic esters of pentavalent phosphorus acids, and the picture becomes correspondingly more straightforward with the role of the generalised anomeric effect clearly apparent.

Most of the evidence concerns the conformational preferences of six-membered cyclic phosphate esters and their derivatives. The phenyl phosphate (*62*, X = OPh) is present in solution in $CDCl_3$ almost exclusively (> 95%) as form (*62a*), with the phenoxy-group axial [115], and the same conformation

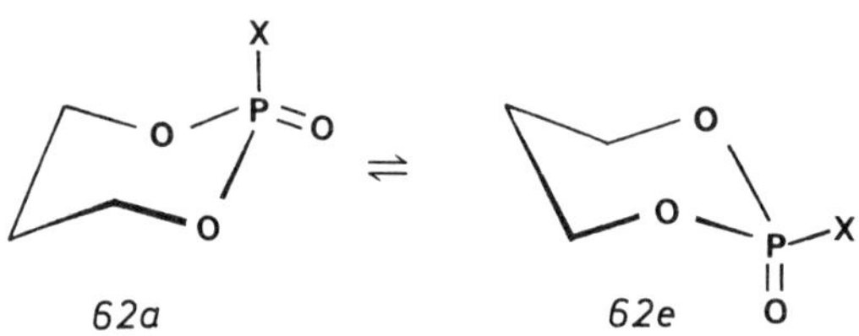

62a 62e

is found in the crystal [116]. An axial preference appears to be general for triesters of this structure, including (*63–65*) [117–120], though this is not so

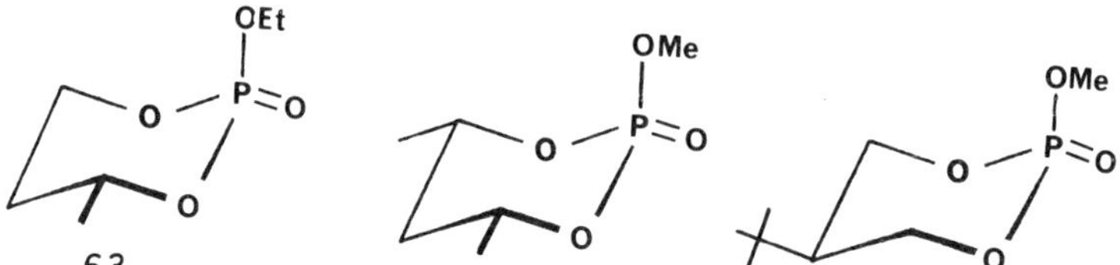

63 64 65

strong that it cannot be overridden by the equatorial (steric) preference of the *t*-butyl group (much less in this ring system than in a cyclohexane): as shown by the preferred conformation (*66e*) of the *trans* isomer of (*65*) [119] in solution. The balance of advantage is evidently a fine one, however, since in the

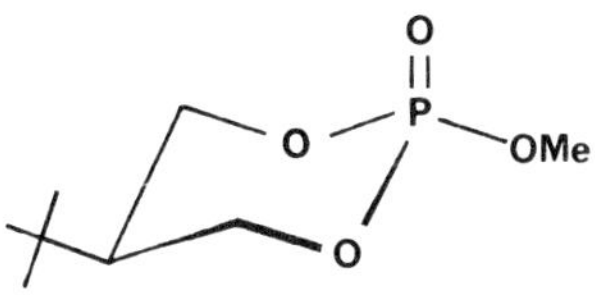

66e

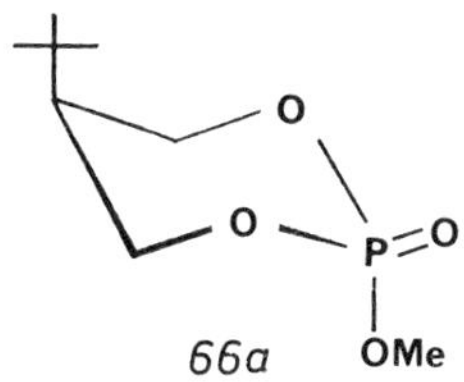

66a

crystal this ester is found to be in the conformation (*66a*) with both methoxy and *t*-butyl groups axial [121]. Estimates of the magnitude of the axial preference are >1.7 [115] and 1.64 kcal mol^{-1} [120] for the phenoxy-group and 1.4 kcal mol^{-1} for OMe [119, 120].

Less electronegative substituents at P show a preference for the equatorial position: or, put another way, a smaller axial preference than phosphoryl oxygen. Hydrogen is axial in (*67a*, R = H), but the methyl group shows

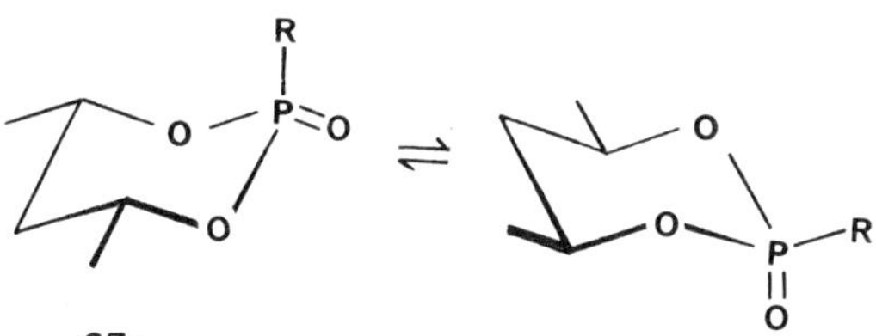

67a 67e

a small preference for the equatorial conformation (*67e*, R = Me) [118], as does phenyl in (*68* R = Ph [115, 120]), which is preferred to the axial isomer by 0.65 kcal mol^{-1} [120]. And crystalline (*68b*), as expected, has the *t*-butyl group at position 5 (rather than the one on P) in the axial conformation [121].

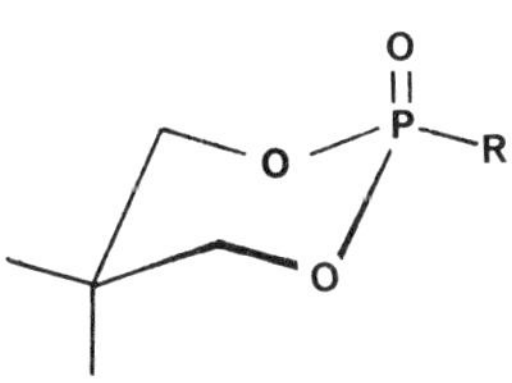

68

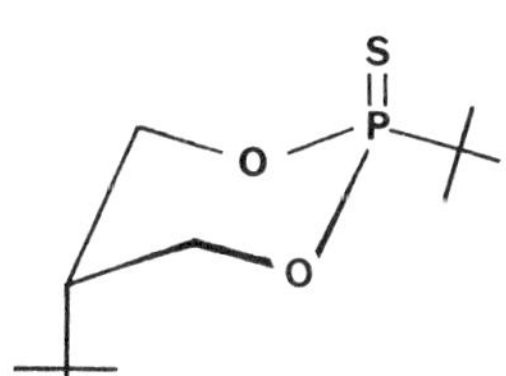

68b

Finally, the dimethylamino-group shows its usual strong preference for the equatorial position in (*67e*, R = Me_2N) [118] and in (*68*, R = Me_2N) [120]. In the latter compound it is estimated that the equatorial isomer is favoured over the axial by 1.11 kcal mol^{-1} [120]. And the conformation about the P–N bond, with the N–R groups approximately gauche to the ring P–O bond, found in the crystal structure of the anti-cancer drug cyclophosphamide (*69*) [122] is consistent with the operation of an exo-anomeric effect in this system.

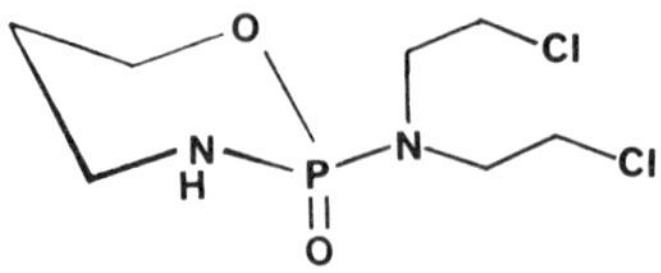

69

Good evidence that the generalised anomeric effect is responsible for these conformational preferences comes from cases where two preferences are opposed in the same molecule. Potentially frustrated conformational preferences are most commonly relieved not by ring inversion (as happens with ester (*66*) above [121], but by conversion to a compromise non-chair conformation (see above, page 12). Thus the twist-boat is the preferred conformation of a number of cyclic phosphate esters which have substituents with conflicting preferences. Examples are the phosphoramidates (*70*) [123] and (*71*) [124], which both exist mainly as the flexible form in solution; as do the phosphate ester analogues [124, 125]. The bicyclic phosphate triester corresponding to (*71*) (Ar = 2,4-dinitrophenyl), for example exists in solution in the twist-boat conformation; while esters of less acidic phenols (Ar = phenyl and *p*-methoxy and *p*-nitrophenyl) appear to be mixtures of chair and flexible forms [124].

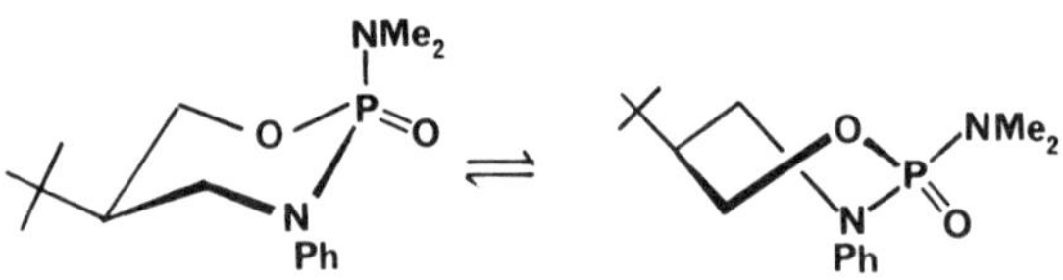

70

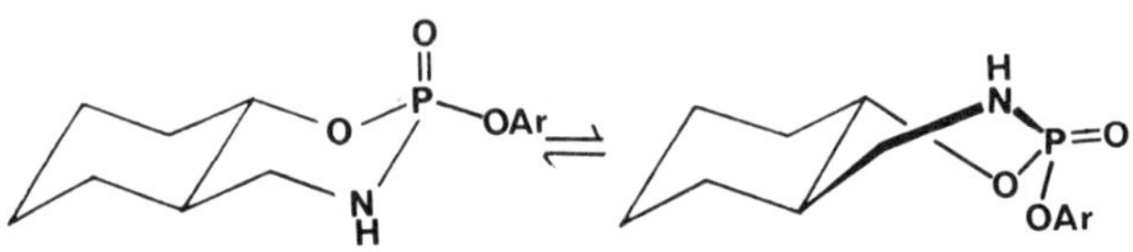

71

A similar case is the dithioester (*72*), where the equatorial preference of the P-phenyl group is strong enough to bring the twist-boat (*72b*) within 0.5 kcal mol^{-1} of the chair in solution [126]. The trans isomer (*73*) exists in the expected, stable chair conformation in solution. But, unexpectedly, *both* (*72*) and (*73*) crystallise as the twist-boat forms [126].

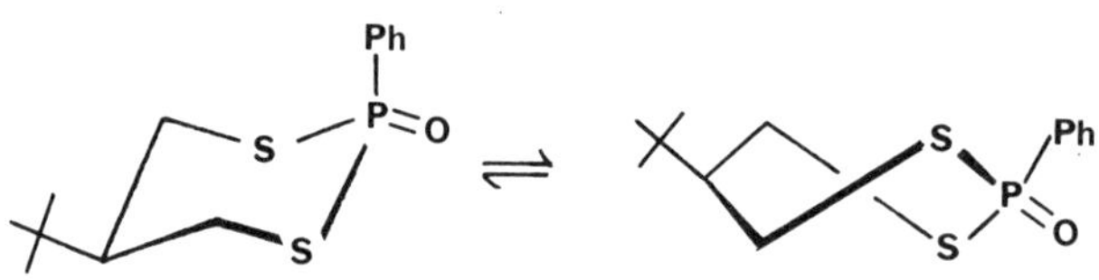

72 72b

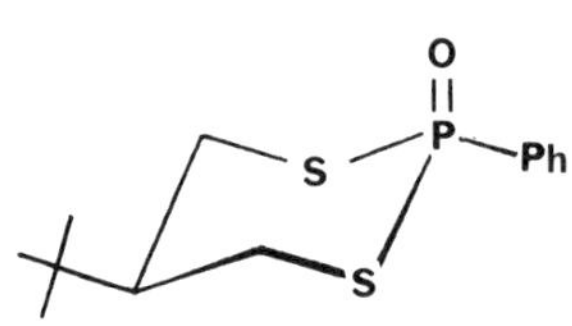

73

c. Transmission through Sulphur

The conformational preferences of sulphur heterocycles show obvious similarities to those of the related phosphorus compounds discussed above. The S–Me group in the S-methylthianium system (*74*) shows only a very small preference (0.275 kcal mol^{-1}) [128] for the equatorial position, while the oxygen atom of the corresponding sulphoxide (*75*) shows a correspondingly weak preference for the axial conformation [5b]; though this may be reversed when

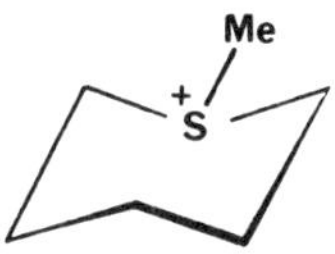

74

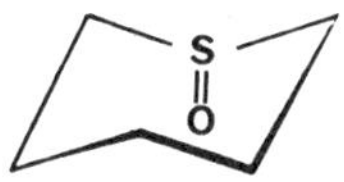

75

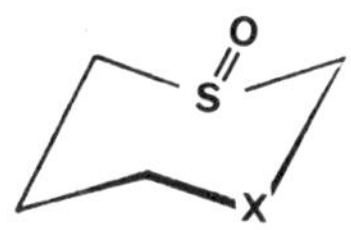

76

electronegative groups (X = O, S, but not C=O) [129] are β to the sulphoxide centre (*76*).

As soon as the basic structural requirements for the generalised anomeric effect are met, marked axial preferences are observed. These are clear-cut in cyclic sulphinate esters (*77*) [130, 131] and strong in cyclic sulphites (*78*)

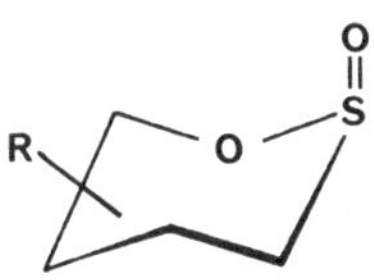

77

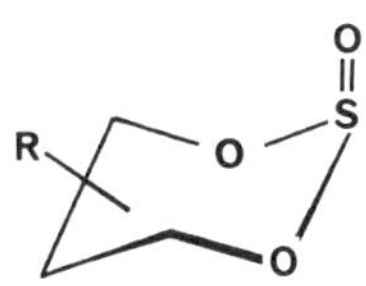

78

Thus equilibration of (*79*) shows that the diaxial isomer (*79a*) is favoured by 1.4 kcal mol^{-1} at ambient temperature [132], so that the generalised anomeric

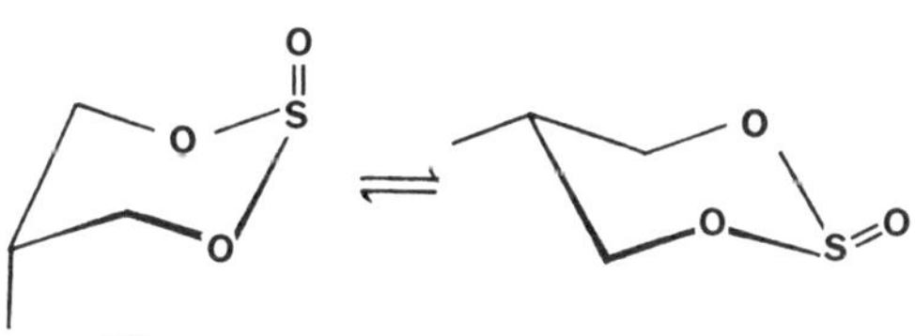

79a 79e

effect, the sum of this and the free energy associated with the equatorial preference of the methyl group, must be at least 2–3 kcal mol^{-1} (less any vicinal lone-pair-lone-pair contribution as discussed below (page 33)). Values in this region have been estimated by other authors. Van Woerden and Havinga [133] found both isomers of (*80*) present in solution, allowing an estimate of **3.5** ± 1 kcal mol^{-1} for the axial preference of the S=O group, within experimental error of an estimate of 2.1 kcal mol^{-1} from similar measurements by Canadian workers [134, 135]. Otherwise measurements on many alkyl substi-

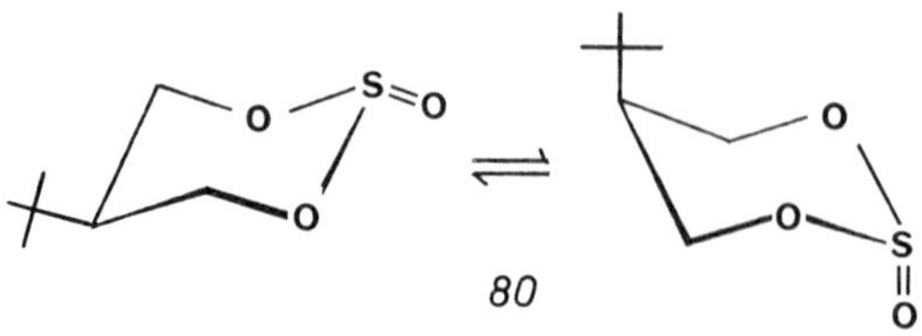

80

tuted sulphites (*78*) show a strong preference for the conformation with the S=O group axial [132–5], which is found also in the (low temperature) crystal structure of the parent compound (*78*, R = H) [136].

Very little information is available about the conformational behaviour of the highly reactive alkyl sulphonate and sulphate esters. Dipole moment measurements on several sulphonates [137a] and sulphinates R_1−SO−OR_2 [137b] gave results consistent with the expected *gauche* conformation (*81*):

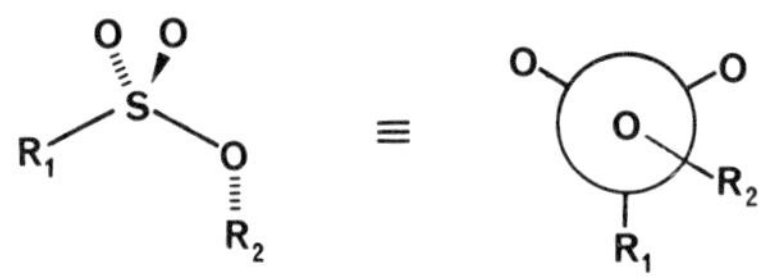

81

but there is a report [138] that methyl methanesulphonate adopts the *trans* conformation: and the only measurements on dimethyl sulphate have been interpreted in terms of the *trans-cis* or *trans-gauche* [139] or even *trans-trans* [140] conformations.

6. The Gauche Effect

The (generalised) anomeric and exo-anomeric effects describe the preference of acetals and related compounds RO−C−X for gauche (*synclinal*) conformations about the C−O bond (see above, page 18). These are not the only systems to prefer gauche to *antiperiplanar* conformations: there exists a similar effect, comparable in magnitude, in systems R−X−Y−R′, where two heteroatoms are bonded directly to each other; and a similar tendency also about the central C−C bond of compounds X−C−C−Y (where X and Y are electronegative groups). Thus the preferred conformation of 1,2-difluoroethane in the liquid phase is (*82*) [2], while that of H_2S_2 is (*83*), with a dihedral angle close to 90° [140].

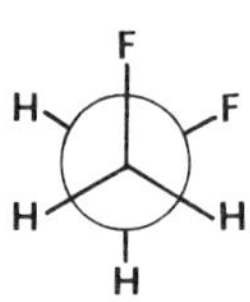

82

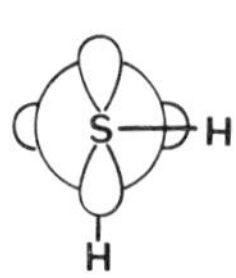

83

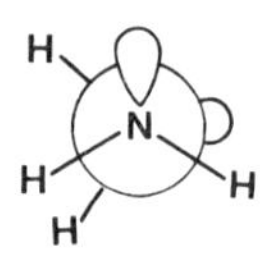

84

Systems of this sort have been discussed by Wolfe [141, 143], who attempted to define a 'gauche effect' as "a tendency to adopt that structure which has the maximum number of gauche interactions between ... adjacent (non-bonding) electron-pairs and/or polar bonds" [141]. The definitive interaction in (*83*), and particularly in hydrazine (*84*), is thus between the lone pair orbitals, which have a marked preference for conformations in which they are perpendicular (as in *83* and *84*) rather than antiperiplanar. Many examples of systems with X–Y bonds, where X, Y are N, P, O, S, Se and As, are given in reference [141].

The most detailed studies on this type of system, using a battery of modern methods, are those of Nelsen [142] on cyclic tetraalkyl hydrazines related to (*85*).

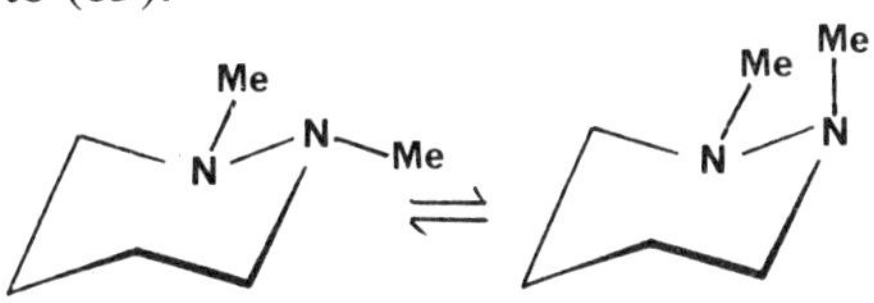

85ee *85ae*

Although the diequatorial isomer (*85ee*) is expected to be considerably more stable on steric grounds, it predominates at equilibrium by only 1.6 : 1[2], with a significant stereoelectronic effect favouring the axial-equatorial form (*85ae*), with gauche lone pairs.

Thus a preference for gauche conformations is rather general when two heteroatoms are sufficiently close to each other within the same molecule, and there have been several attempts to find a common basis for this preference. Wolfe's original definition, quoted above, important as a landmark on the way to a more complete understanding of these effects, turns out to be of limited applicability. Systems with adjacent heteroatoms do prefer conformations with perpendicular lone pairs, as discussed above, but the definition does not predict correctly the relative stabilities of the gauche and *antiperiplanar* conformations of most 1,2-disubstituted ethanes, XCH_2CH_2Y [144], since only a small number of compounds, with strongly electronegative substituents X and Y, actually prefer the gauche conformation. Thus the definition holds generally for lone pair-lone pair systems, but only in special cases for polar bond-polar bond interactions. Most seriously, though, it fails completely for lone pair-polar bond interactions. These are precisely the interactions which give rise to the anomeric effect, and though the anomeric and exo-anomeric effects certainly favour gauche conformations, (e.g. *86* and *87* for FCH_2OH [141] and $MeOCH_2Cl$ [145] respectively) the polar bonds have the minimum rather than the maximum number of gauche interactions with non-bonding lone pairs. Thus this type of system shows a strong preference for conformations with lone pairs and adjacent polar bonds *antiperiplanar*; in contrast with hydrazines and peroxides, where the *antiperiplanar* arrangement

[2]Remarkably, the N.N-diethyl compound exists exclusively in the ae conformation [142].

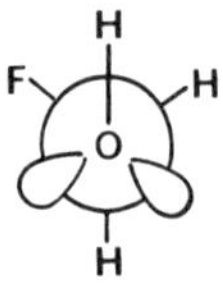

86

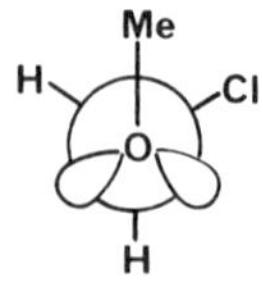

87

of vicinal lone pairs is clearly unfavourable. Evidently any theory which treats lone pairs and polar bonds as equivalent in any sense is likely to break down at this point.

Recent work from the laboratories of Eliel and Zefirov on X/\/Y compounds has underlined the complexity of this sort of system, where steric, stereoelectronic and dipole-dipole interactions may all be of the same order of magnitude, making generalisations difficult. Phillips and Wray [145] stressed that the more electronegative are X and Y, the more the gauche conformation is favoured (and found a simple correlation between the free energy difference between the gauche and *antiperiplanar* forms and the sum of the electronegativities of X and Y). The same effect is apparent in the behaviour of the 5-halogen-1,3-dioxanes (*88*), where the preference for the equatorial conformation (*88e*) decreases markedly in the series X = I > Br > Cl > F [146, 147], to such effect that the axial isomer (*88a*), which has the gauche

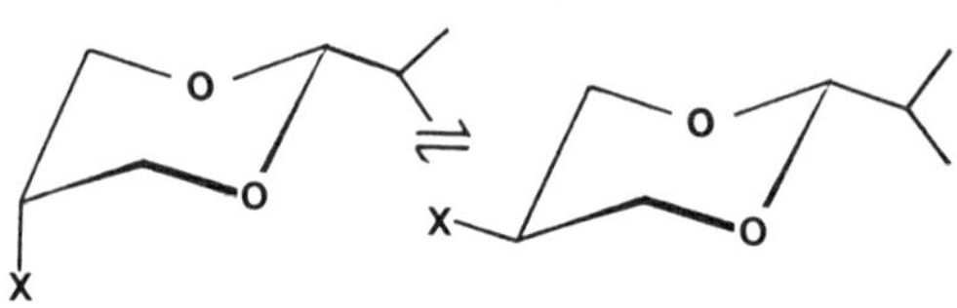

88a 88e

conformation about the two O−C−C−X bonds, is favoured for the fluorocompound in all solvents [146]. A similar axial (gauche) preference for $MeSO_2$ > MeSO > MeS (the latter group being mainly equatorial) is consistent with this being a stereoelectronic rather than a steric or dipole-dipole effect. (Other partially or fully positively charged groups (NO_2, NMe_3^+, SMe_2^+) also favour the axial position, and the authors suggest that electrostatic attraction for the axial lone pairs of the ring oxygens may be another factor operating in this system.)

Detailed conformational analysis of this (*88*), the related dithiane system (*89*) [148, 149], and of several *trans*-1,2-disubstituted cyclohexanes (*90*) [150]

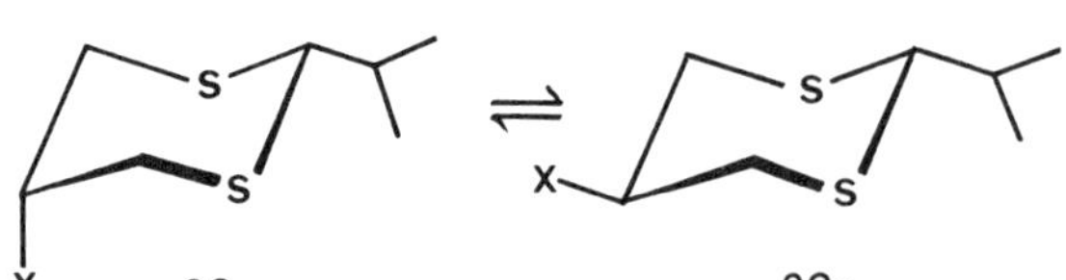

89a 89e

has led to the identification of a 'repulsive gauche effect'; that is to say an

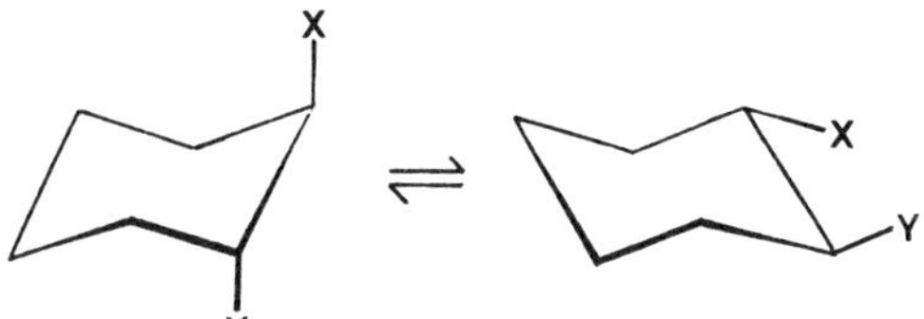

90aa *90ee*

effect which positively disfavours the gauche relationship (as in *88a, 89a* and *90ee*) for certain groups X and Y, and thus operates in the opposite sense to the original 'gauche effect'; which now has to be specified as the 'attractive gauche effect'. As is to be expected, the 'attractive gauche effect' is found for the most electronegative groups, specifically for vicinal OR/OR and F/I interactions [86, 149]. For a range of intermediate cases X/Y = MeO/Cl, MeO/Ar, MeO/I in (*90*), probably O/SMe in *88*, X = SMe) the observed equilibrium constant for the interconversion of the two forms of (*88, 99* or *90*) is close to the best calculated value, so that no significant gauche effect is apparent. But for large groups (S/Cl, S/Br, S/S and S/Br in (*90*), X = OMe, SMe in *89*) the *antiperiplanar* conformation (*89e, 90aa*) is clearly favoured more strongly than expected on the basis of classical steric and polar interactions [148–150]. The explanation of this 'repulsive gauche effect' currently favoured by both groups is in terms of non-bonded repulsion between the 3p orbitals of the large substituents X and Y, which is absent when they are *antiperiplanar*, maximal when they are *synperiplanar* (*91*), and still significant when they are gauche (*synclinal*) to each other. (This repulsive interaction

91

was originally called the 'hockey sticks effect' by Zefirov [150a], for obvious reasons (*91*).

Thus the stereoelectronic effect of interest, the 'attractive gauche effect' is not a major effect, but it is significant in systems X−C−C−Y where X and Y are strongly electronegative groups (F, O), where it favours the gauche conformation in spite of the opposing forces – steric effects and dipole-dipole interactions – favouring the antiperiplanar form.[3]

---

[3] The effect probably also contributes to the relative instability of cyclic systems containing the O−C−C−O unit fixed in the *antiperiplanar* conformation. Examples are 3',5'-cyclic nucleotides, like cyclic AMP (*i*), which have unusually exothermic heats of hydrolysis, attributed in part to the *antiperiplanar* arrangement of the furanose $O_1'-C_5'$ and the two C−O bonds of the six-membered ring (a monocyclic analogue exists in the conformation (*ii*) with the OMe group axial) [151a]: and the celebrated *trans-bis*-acetal (*iii*) of glyoxal with ethylene glycol, which has *antiperiplanar* arrangements across both O−C−O−C and both O−C−C−O systems, and is thus destabilised by both anomeric and gauche effects. It is probably no coincidence that (*iii*) has never been prepared, despite numerous attempts [151b]. (The known *cis*-isomer, (*iv*) appears to have an unusually low barrier to ring inversion.)

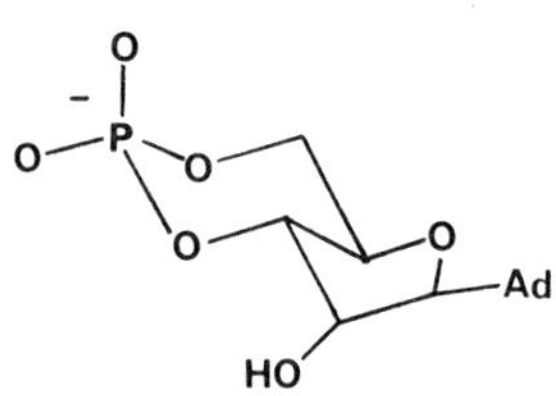

(i)

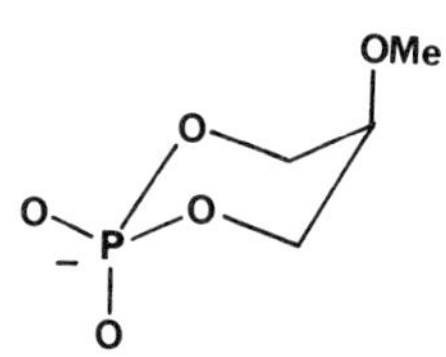

(ii)

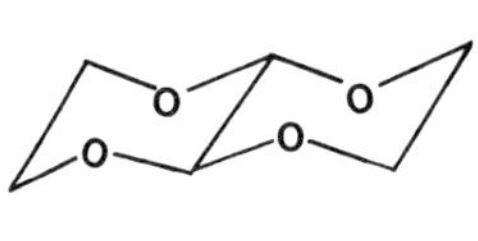

(iii)

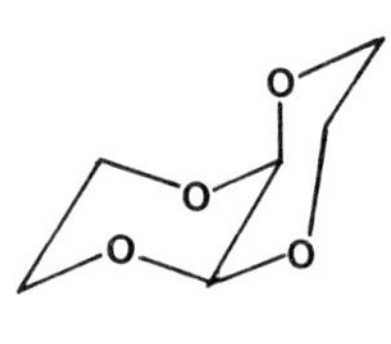

(iv)

B. The Electronic Basis of the Effects

1. Introduction

Throughout our discussion in section A of the experimental evidence for the various manifestations of the anomeric effect, and the preference for gauche conformations in systems with adjacent or vicinal heteroatoms, we have been concerned with stereoelectronic effects on conformation: that is to say, orbital-orbital interactions with well-defined geometrical requirements which are clearly different from, and frequently override, the non-bonded repulsions between orbitals which are the basis of steric effects. If there is a common basis for all of these effects, therefore, it must lie in orbital-orbital interactions of this sort.

The effects on conformation result specifically from restrictions on rotation about the central bonds of systems X–C–C–Y, C–X–Y–C and C–X–C–Y over and above the usual threefold barriers expected for rotation about single bonds; and are clearly most marked for those systems (C–X–Y–C and C–X–C–Y) in which at one or both ends of the central bond is a heteroatom with one or two non-bonded electron-pairs. The new factor is thus likely to be the presence of these non-bonding orbitals, and the discussion of the effects has centred on their interactions with other non-bonding or bonding orbitals. The conformational preferences of tetraalkylhydrazines, for example (page 33) could be rationalised on the basis that the conformation (e.g. *85ee*) with *antiperiplanar* lone pairs is significantly destabilised, for stereoelectronic reasons, compared with the isomer (*85ae*) in which they are close to gauche *(synclinal)* to each other [142].

The preference for gauche conformations which is the basis of the (generalised) anomeric and exo-anomeric effects can also be expressed in terms of a preferred orientation of localised non-bonding electron-pairs. In the preferred axial conformation of a tetrahydropyranyl acetal (*92a*) one of the lone pairs on the ring oxygen atom is *antiperiplanar* to the exocyclic

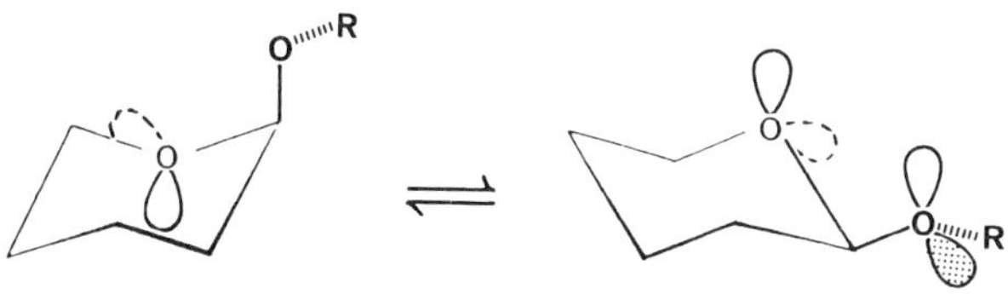

92a *92e*

C–O bond. In the equatorial isomer (*92e*) in contrast a bonding orbital is *antiperiplanar* to this bond.

Similarly the preferred conformations about the exocyclic C–O bond of the acetal group in such systems (A1 in *92a*, E1 in *92e*), which are controlled by the exo-anomeric effect (see page 18), are conformations with lone pairs on the exocyclic oxygen *antiperiplanar* to the endocyclic C–O bond of the acetal group.

This is quite general. The conformation about the O–C bond of a system R–O–C–X is phenomenologically most simply described in terms of the dihedral angle between the R–O and C–X bonds (*93*), but the favoured

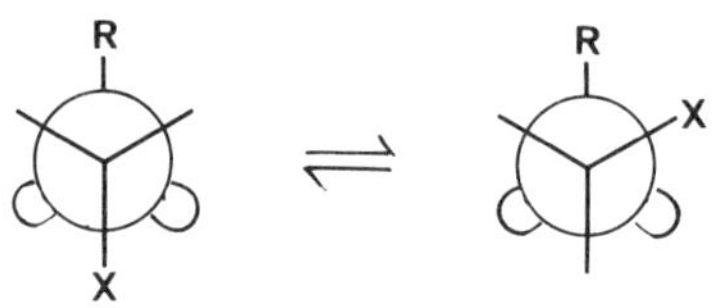

93ap *93g*

gauche conformation (*93g*) always has a non-bonding electron pair *antiperiplanar* to the bond to the electronegative substituent X, and evidence has been accumulating that this is a key feature.

Various rationalisations of the anomeric effect have been offered over the years. Edward [14] originally suggested that electrostatic repulsive forces between non-bonding electron-pairs and the C–X bond (maximal in conformations (*93ap*) and the corresponding (*92e*) account for the preference for (*93a*) and (*92a*); but this and related explanations in terms of dipole-dipole interactions fail to account quantitatively for observed axial preferences, particularly in halogeno-ethers [96], and do not predict the bond-length changes which are characteristic of systems showing the anomeric effect (see below, page 52). Nor does the explanation in terms of electrostatic repulsion between parallel or *syn*-axial lone pairs on geminal electronegative atoms, as in the unfavourable conformations (*94, 95*) of 1,3-triazines or dimethoxymethane, which was christened the rabbit-ear effect [92].

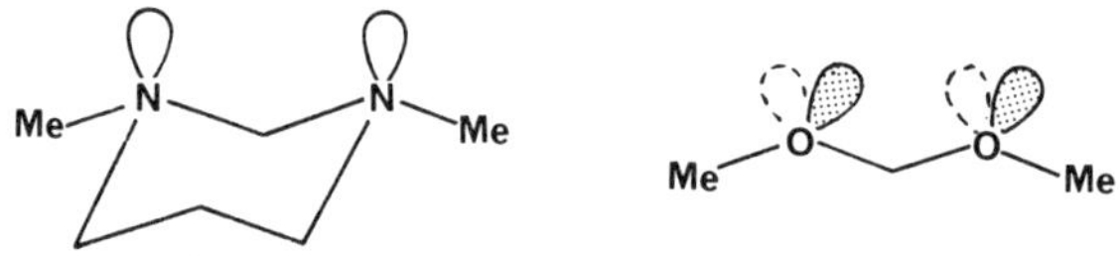

94 *95*

The first 'explanation' which is still accepted as being substantially correct, though incomplete, arose from the structural work of the Leiden school on α-halogenoethers. It was shown that, associated with the preference of methoxymethyl chloride for the gauche conformation [145], or of *trans*-2,3-dihalogeno-1,4-dioxanes for all-axial conformations (*96*) [96], is a charac-

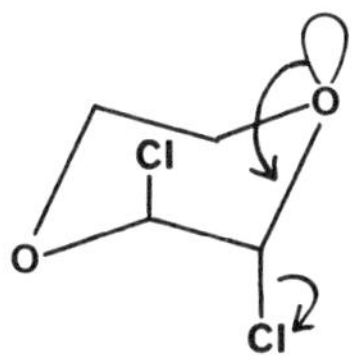

96

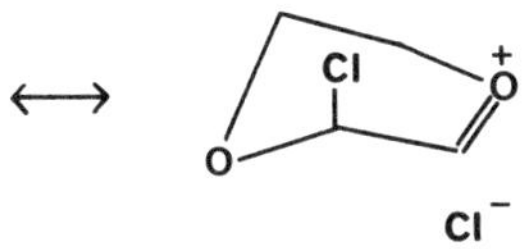

97

teristic lengthening of the C−X bond, and a concomitant shortening of C−O, in the system R−O−C−X [96, 152]. This was ascribed to mixing of the non-bonding electron pair orbital on oxygen with the antibonding orbital of the adjacent C−X bond (see below, page 64, for further discussion of this interaction); what in valence-bond terminology would be described as double-bond no-bond resonance (*96*↔*97*) [5c]. It is only in special cases that this particular pattern of bond length changes is clearly apparent (for a full discussion of this question see below, page 52), but it is clearly associated with a particular conformational arrangement (*96*), and thus, by implication, with the anomeric effect. So that any explanation of the anomeric affect should explain these bond length changes also.

By this point it has also become abundantly clear that any explanation of the anomeric and related effects must involve the non-bonding electron pairs of oxygen (or other hetero-atoms) involved. So the properties of those electron pairs become a crucial element in the discussion. We have depicted them so far in the way most organic chemists recognise, as localised sp^3-hybrid orbitals on tetrahedral centres, but this is undoubtedly a simplification; so it becomes important to decide whether the simplification involved is justified, unjustified but harmless, or actually misleading. There is a growing body of physical evidence about non-bonding electron pairs, which should eventually lead to clear-cut answers to such questions. Thus far the evidence is indicative, rather than compelling, and we need to look rather closely at what is known about the status of lone pair electrons before taking the argument further.

2. The Status of Lone Pair Electrons

Unless they have a particular reason not to, chemists generally visualise the electronic structure of molecules in terms of localised molecular orbitals, corresponding to the bonds between the atoms that they write on paper. A rigorous representation of bonding in the ammonium ion, or the isoelectronic methane, involves a combination of the four 1s orbitals of the four hydrogen atoms with the four (2s, $2p_x$, $2p_y$ and $2p_z$) valence orbitals of the central nitrogen or carbon atom [153]. The geometry of the four resulting bonding molecular orbitals (three equivalent, the fourth involving the 2s orbital overlapping with all four hydrogen orbitals) does of course lead to the tetrahedral arrangement of the ligands (Fig. 1), but not in a way that is immediately obvious.

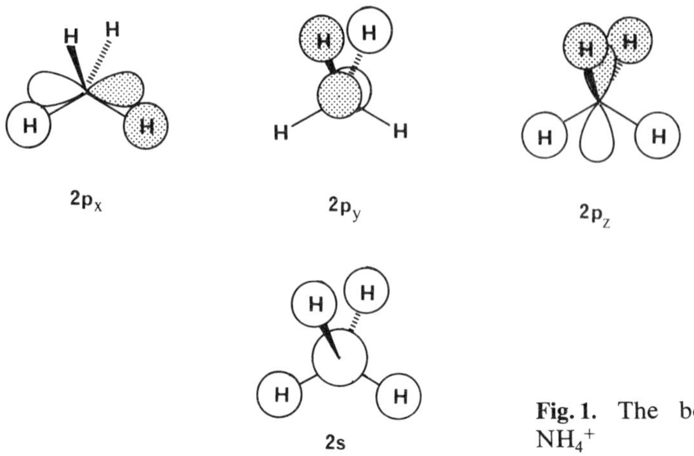

Fig. 1. The bonding orbitals of NH_4^+

The description in terms of localised (sp^3-hybrid) molecular orbitals, on the other hand, has the tetrahedral geometry built in, and relates specifically to the bonding to individual hydrogen atoms. Furthermore, it leads not only to a readily understood representation of the structure of CH_4 or NH_4^+, but also to the simplest prediction of what happens when the ammonium ion loses a proton to give ammonia.

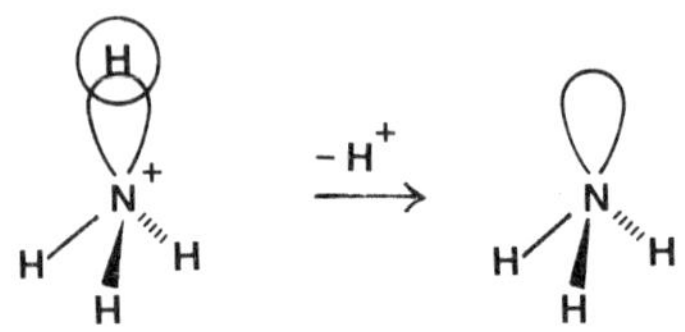

In principle it is possible for nitrogen to use its three p orbitals to form three *pσ*-bonds, each at right angles to the other two. (This picture, leaving the low-lying s-orbital untouched, gives a good account of what happens in the third row of the periodic table, where PH_3 is practically non-basic, and has HPH angles close to 90° [154]. In practice it is clear that the 2s orbital is involved in bonding, because the HNH angles of ammonia are not 90°, but 107°, close to the tetrahedral angle of 109° 28′. Both this geometry and the basicity of ammonia indicate the presence of electron density (effectively a fourth ligand) in the region of the axis of symmetry. Nor can it be described as a simple 2p-orbital, since this would imply planar geometry, with HNH angles, presumably, of 120°. In fact a prediction based on the NH_4^+ structure, with a proton removed from one of the N–H bonding orbitals to leave a non-bonding pair of electrons, provides a rather good description of the ammonia molecule (for a discussion of the structure in terms of canonical molecular orbitals see Reference 154a).

Thus arises the picture of the non-bonded electron pair (lone pair) of electrons as a localised orbital with characteristics such as steric requirements and directionality similar to those of other, bonding, orbitals. The directionality is that imposed by the prevalent hybridisation (geometry) at the centre concerned, and for ammonia and symmetrical tertiary amines simply lies along the C_{3v} symmetry axis. The steric requirements ("size") of the orbital depend on the charge type and electronegativity of the nucleus: some relevant calculations [155] show how this is expected to increase with increasing charge in the series $NH_3 < NH_2^- < NH^{2-} < NH^{3-}$.

The electrons on more electronegative atoms, such as oxygen and fluorine, are under the influence of a larger nuclear positive charge, so that lone pair orbitals in particular are expected to be held more tightly, and thus to be less diffuse (smaller in size) in the series N > O > F. The question of directionality, essentially unambiguous for non-bonding electron pairs on nitrogen, also becomes more complicated. There is little that can usefully be said about the directionality of lone pairs on halogen atoms, but the properties of the non-bonded electron-pairs on oxygen hold the key to the stereoelectronic effects discussed in this book.

The bond angle at oxygen in water, alcohols and acyclic ethers generally lies in the range $104 \pm 1°$, still much closer to the tetrahedral angle than to 90°. The two non-bonded electron-pairs are formally described in terms of a unique pair of canonical molecular orbitals, one with π-symmetry, one with σ-symmetry, as shown in Fig. 2a; or alternatively in terms of any one of an

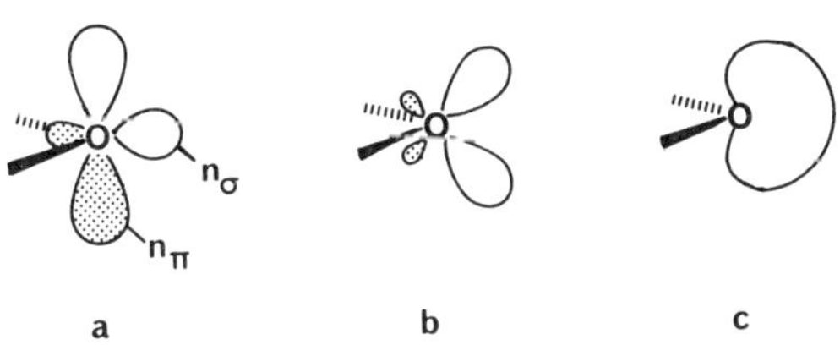

Fig. 2 a–c. Different representations of the non-bonding electrons of bicovalent oxygen. **a** Canonical and **b** sp^3-hybridised orbitals; both give rise to the same electron-density distribution (**c**)

infinite number of sets of localised molecular orbitals, of which the pair shown in Fig. 2(b) are the most familiar, and represent the closest approach to tetrahedral symmetry. It is important to understand that (a) and (b) of Fig. 2 are mathematically equivalent representations of the non-bonding orbitals of the oxygen atom. In particular, when the individual wave functions are squared and added together (Fig. 3) [156] they give rise to the same electron-density distribution, shown in Figs. 2(c) and 3(c).

It is this kidney-shaped electron-density distribution which corresponds to physical reality, and is responsible, for example, for the van der Waals radius of oxygen in the direction away from the two ligands. To analyse orbital-orbital interactions within the molecule, on the other hand, it is necessary to consider the individual molecular orbitals of Fig. 2(a) or 2(b). For many purposes either representation may be used, and organic chemists generally prefer the picture [2(b)] in terms of hybrid orbitals. But where the symmetry of the interaction, or the energy of the non-bonding electrons is

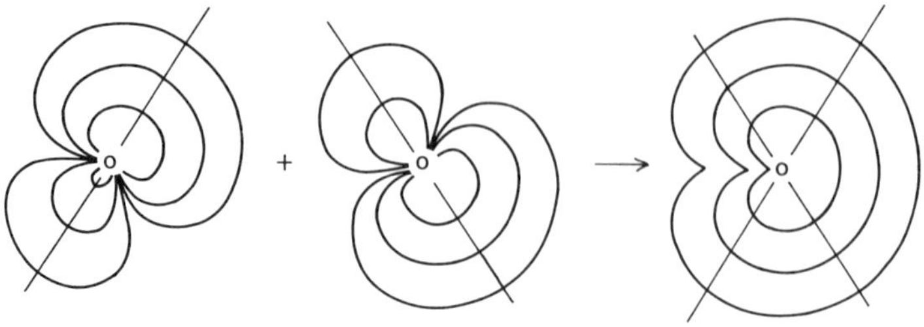

Fig. 3a–c. Calculated electron-density plots for the non-bonding electron pairs of bicovalent oxygen. **a** and **b** represent electron densities for individual hybrid orbitals, **c** shows the sum of the densities for the two lone pairs [156]

crucial, it is generally considered necessary[4] to use the canonical molecular orbitals of Fig. 2(a) [157]. These differ substantially in energy, the σ-type orbital (n_σ, formally sp) lying below the π-type orbital (n_π, formally a 2p orbital of oxygen) by 1–2 ev (20–45 kcal mol^{-1}) [158, 159]. In saturated oxygen-containing molecules the highest occupied molecular orbital (HOMO) is generally a non-bonding orbital on oxygen, so the origin of the electron-donor capability of two-coordinate oxygen, whether inter- or intramolecular, is the non-bonding electron-pair, n_π, with π symmetry.

Experimental evidence about the electron density in the lone pair region is generally consistent with the picture deduced from theoretical considerations (Figs. 2(c), 3(c)). Allinger and Chung [160], for example, found that they could not successfully extend their molecular mechanics calculations to 5-alkyl-1,3-dioxanes without explicitly including lone pairs. Within the force-field used, and assuming approximate sp^3-hybridisation as in Fig. 2(b), the parameters which worked best put centres of electron density at 0.5 Å from the nucleus, with an angle of 140° between them. This result does not prove that there are sp^3-type lone-pairs on oxygen – it is obviously indirect, and equally consistent with the distribution shown in Fig. 2(c) – but it does suggest that electron density arising from the presence of non-bonding electron pairs is a physical reality.

a. Evidence from Hydrogen-bonding

So, in a more direct way, does hydrogen-bonding. A hydrogen-bond acceptor is normally a system with lone pair electrons, and since all hydrogen bonds have an electrostatic component their existence is *prima facie* evidence for a significant electron density in the lone pair region. Much more important, however, is the possibility of making deductions about the shape of the elec-

[4] For further discussion of this point see page 78 below.

tron density distribution. Hydrogen bonds are generally close to linear. If the H-bond points to the centre of electron-density of the lone pair (and the interaction is not large enough to perturb the non-bonding orbital seriously) then we may expect the vector $Y..\vec{H}-X$ to define the directionality of the lone pair orbital.

Much structural information is consistent with this idea. The H−F..H angle in crystalline HF is 120°, so that the chain is not linear but zig-zag [161]; whereas the chains in crystalline H−C≡N..H−CN are linear [162]. Many H-bonds to C=O are in the plane of the carbonyl group, with C=O..H angles close to 120°. And the three-dimensional structure of ice, with each oxygen atom surrounded by four hydrogens, two bonded and two H-bonded, in perfect tetrahedral symmetry, has probably influenced ideas on lone pair directionality more than any other single piece of evidence [163, 164].

In all of these cases the structural information is consistent with well-developed lone pair orbitals, having marked directionality, which is moreover that expected for sp, sp^2 or sp^3-hybridisation. However, it should be said at once that this simple picture, though not uncommon, is very far from being the rule. The three-dimensional arrangement of atoms in a crystalline solid is subject to strong packing forces, which often dictate the direction of hydrogen bonds. And we have no *a priori* way of knowing what corresponds to the 'natural' directionality – the cases with the 'geometry' expected for simple hybrid lone pairs, or the wide range of exceptions. The best that can be done is to analyse H-bond directionality in as large a number of structures as possible, in the hope that any systematic, "natural" preference will show up against the presumably random background of crystal packing forces.

Studies of this sort are available for hydrogen bonding involving carbonyl and hydroxyl groups as acceptors. In the case of the carbonyl group simple hybridisation theory predicts that oxygen will have two lone pairs at the trigonal angle of 120°, in the plane defined by the three bonds to carbon (*98*).

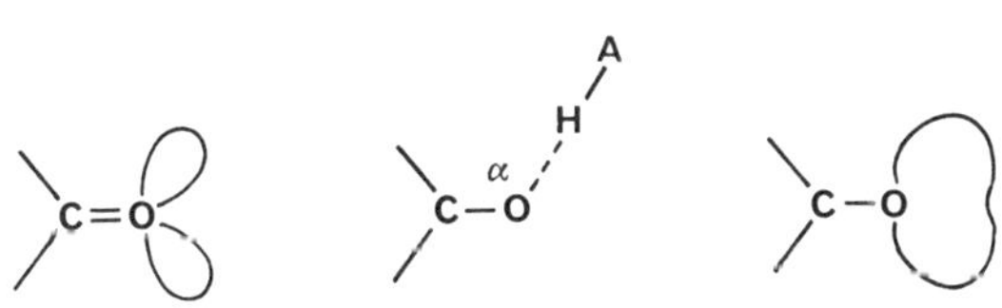

98 *99* *100*

The expected C=O..H angle (α in *99*) is thus 120°. An early study by Donohue [165] showed that hydrogen bonds to the carbonyl oxygen of many amides and ureas make angles α very far from 120°, and within a small (and biassed) sample found no evidence for a significant angular preference over the range $\alpha = 120$–170°, nor an absolute preference for the H-bond to lie in the plane of the carbonyl group. But more extensive recent results are consistent with the idea of a preferred directionality. A systematic analysis of over 80 structures determined by neutron diffraction, which has the advantage that the position of the hydrogen atom can be pin-pointed, also showed a broad range of C=O..H angles (α) for H-bonds to O−H and N−H donors [166]; but the great majority lay in the range $120 \pm 15°$, consistent with an electron density distribution similar to that predicted by the lone pair picture. (Which in the case of the lone pairs of the carbonyl group, which have a wider angular separation than those at sp^3-oxygen, might be expected to look something like (*100*). Furthermore, almost all the short (and many of the long) O−H..O=C bonds lay close to the plane of the carbonyl group. The implication from this study is clearly that there is a preferred directionality of H-bonding to carbonyl oxygen, close to that expected on the basis of trigonal lone pairs, but that H-bonds with a very wide range of geometries can form if crystal packing forces so dictate.

The picture at OH oxygen is slightly different. Two statistical studies of a large number of O−H..O hydrogen bonds taken from X-ray [167] and neutron diffraction [168] measurements are consistent in finding no preference for a particular directionality, which might be associated with two distinct lone pairs. There is, on the other hand, a general preference to form H-bonds in the plane in which the lone pairs are expected to lie, with a mean value

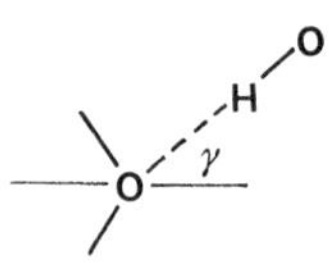

101

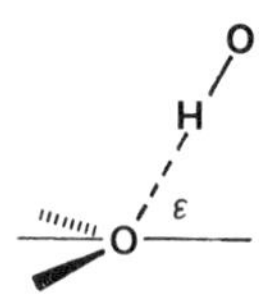

101a

of γ close to zero (*101*) (standard deviation 12–14°) [167, 168]. Within this plane, however, (i.e. the plane of the paper in *101a*) the distribution of data is almost uniform over the range $60° > \varepsilon > -60°$, consistent with a kidney-shaped non-bonded electron density distribution as depicted in Fig. 2(c), above.

When a lone pair takes up a proton, to form a covalent bond, we know that the bond will be formed in the direction expected from hybridisation theory. It is to be expected, therefore, that the stronger the hydrogen bond, the more likely it is to be formed with well-defined preferred directionality. There is little experimental support for this idea [166, 169], but for the most part the results seem to indicate that the (always relatively weak) hydrogen bond causes little perturbation of the electron density distribution.

b. Electron Density Measurements

The most recent, and potentially most important, experimental approach to defining the status of lone pair electrons is the measurement of electron densities by accurate X-ray diffraction techniques [170, 171]. X-rays are diffracted by all the electrons in a crystal, and the contribution from the valence electrons is generally small enough to be ignored in the usual least squares model, which assumes that the crystal is made up of free, spherically symmetrical atoms. (This leads to significant errors in the atomic co-ordinates of the lightest atoms, particularly hydrogen, and these are best determined by neutron diffraction, neutrons being diffracted by nuclei rather than electrons.) On the other hand it is precisely the contribution from the valence electrons which is of interest to the chemist (once he knows the structure), and this can be isolated from the observed electron density map by subtracting the density expected for free symmetrical atoms in their electronic ground states, centred at the known atomic positions in the crystal. The resulting electron density distribution (called the *deformation density* because it represents the deformations which occur when atoms interact to form molecules) often shows features, such as a build-up of density in the region between the nuclei or, in particular, in the lone pair region, consistent with the expected pattern of molecular orbitals. So in principle this technique allows the direct study of the shapes of bonding and non-bonding orbitals.

The results so far are promising but not definitive. When enough very accurate data are available it seems likely that it will be feasible to make firm statements about the distribution of electrons in bonds, and perhaps the lone pair region. But such highly accurate measurements need great care, well-behaved crystal structures, measurements at low temperatures and, preferably, centrosymmetric molecules, and only a relatively small number of such studies are so far available. A much larger number of similar but less accurate measurements can be analysed statistically, perhaps to give similar information; and the general conclusions which can be drawn at present are as follows.

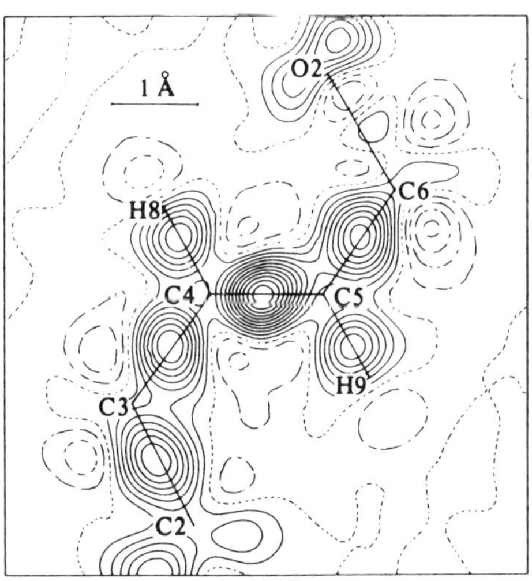

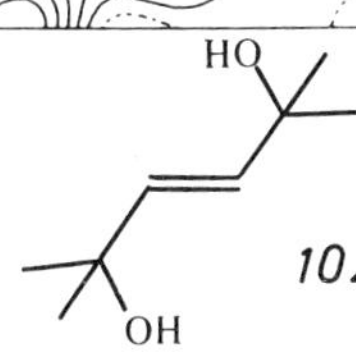

102

Fig. 4. Section of the experimental deformation-density distribution in the plane befined by C(3)C(4)C(5), for 2,5-dimethyl hexene-2,5-diol (*102*), showing positive electron-density in the region expected for σ-bonding orbitals. (Reproduced with permission from H.R. van der Wal and A. Vos, Acta Cryst. *B35*, 1804 (1979).)

More prominent in the deformation density maps are bonding densities in double and triple bonds [172–176], though positive density also shows up clearly in single bonds in accurate studies (see Fig. 4 [177]). Lone pair densities are considerably more diffuse, but are clearly apparent in favourable cases at sp^2-hybridised nitrogen (Fig. 5 [178]) and carbonyl oxygen (Fig. 6 [179–184]).

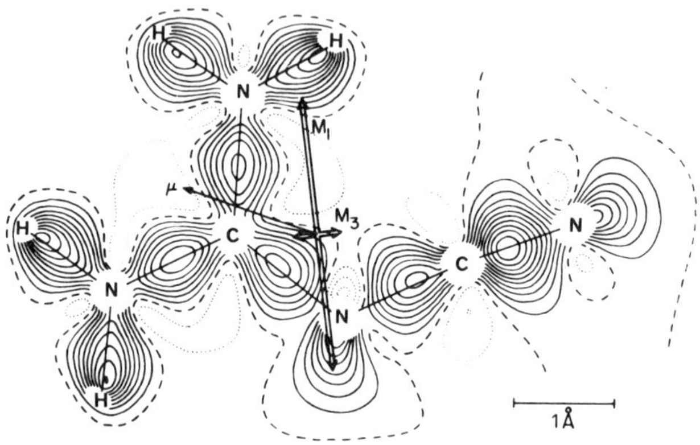

Fig. 5. Experimental deformation-density in the mean molecular plane of 2-cyanoguanidine. Note the well-developed electron density in the nitrogen lone-pair region. (Reproduced with permission from F.L. Hirschfeld and H. Hope, *Acta Cryst. B36,* 406 (1980).)

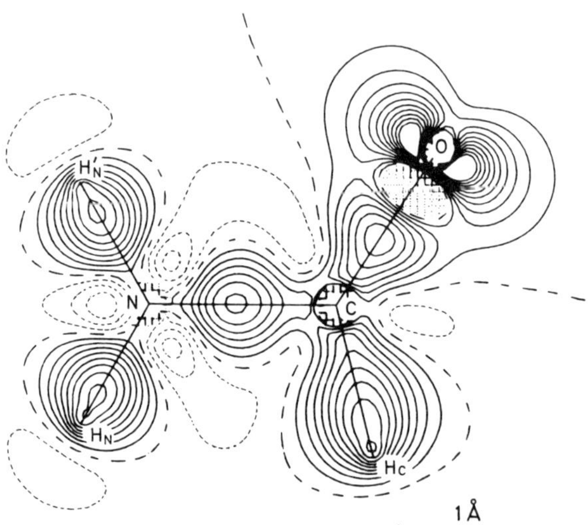

Fig. 6. Deformation density in the mean molecular plane of formamide. Note the well-developed electron density distribution in the lone pair region of the carbonyl oxygen. (Reproduced with permission from M. Eisenstein, Acta Cryst. *B35,* 2614 (1979).)

Most relevant for our discussion, and most elusive, are lone pair densities at saturated oxygen and nitrogen. Lone pair peaks on amine nitrogen atoms are just about visible in a few recent determinations [185, 186], and enough relevant oxygen compounds have been studied to allow some reasonably firm preliminary conclusions.

The only electron-density measurements on an ether (the polyether 1,4,7,10,13,16-hexaoxacyclooctadecane, or 18-crown-6 [186]) show rather diffuse peaks in the lone pair regions, though their spatial distribution may "correspond roughly to what we would expect for sp^3-lone pairs rather than sp^2, p lone pairs" [186]. The acetal and hemiacetal oxygens of sucrose show slightly better resolved 'crescents' of electron-density in the lone pair regions, very much like the electron density around the numerous hydroxyl oxygens of this molecule [187]. It may be that hydroxyl oxygen is intrinsically better suited for deformation density measurements than oxygens in ethers, because the OH group is always likely to be involved in a network of hydrogen bonds, which will restrict the thermal motion that is an important source of uncertainty in such measurements.

Whether this is true, or not, the most detailed pictures currently available of electron density in the lone pair regions of two-coordinate oxygen involve the OH group, in water [188], hydrogen pyroxide [189] and the diol (*102*). In every case the lone pair electrons are engaged in hydrogen bonding, with geometries varying from close to tetrahedral to planar (in the water of crystallisation of oxalic acid [188]); but the resulting perturbations are not large enough to obscure the essential similarity of the electron density distributions – (Figs. 7, 8 and 9).
As expected, (see above, page 41) the two non-bonding electron-pairs give rise to a single, kidney-shaped region of high electron density, with little evidence, except in the *internal* electron density distribution of Fig. 9, for separate constituent orbitals.

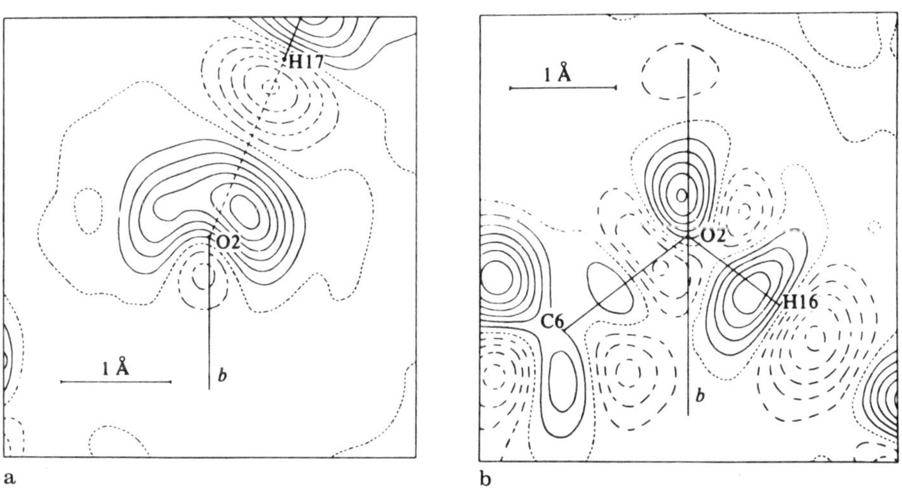

Fig. 7a and b. Deformation densities in the lone pair region for one OH oxygen of the diol *(102)*, **a** perpendicular to and **b** in the C–O–H plane. (Reproduced with permission from H.R. van der Wal and A. Vos, Acta Cryst. *B35*, 1804 (1979).)

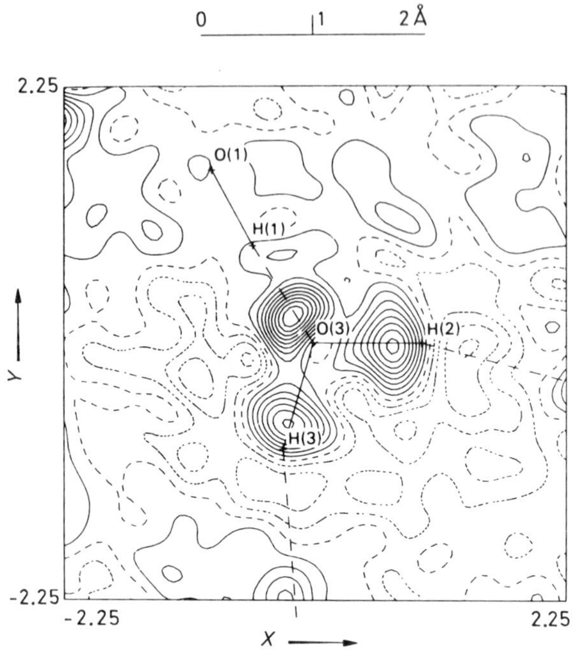

Fig. 8. Deformation density in the lone pair region for the water of crystallisation of oxalic acid, plotted in the plane of the water molecule. (Reproduced with permission from E.D. Stevens and P. Coppens, Acta Cryst. *B36*, 1864 (1980).)

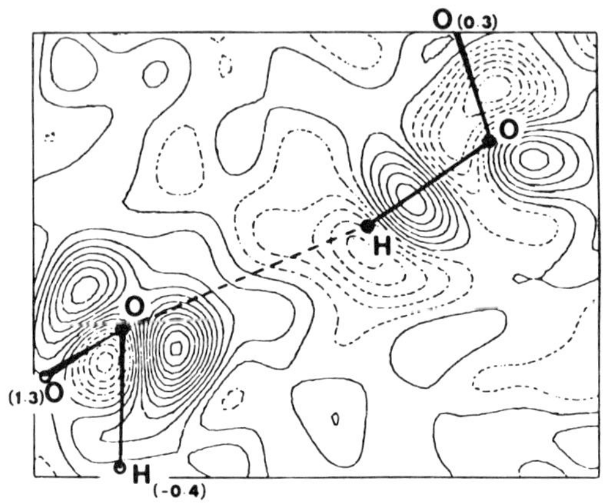

Fig. 9. Deformation density in the region of the hydrogen bond of crystalline H_2O_2. (Reproduced with permission from J.-M. Savariault and M.S. Lehmann, J. Am. Chem. Soc. *102*, 1298 (1980).) Copyright (1980) American Chemical Society.

3. Calculations on X⁀Y Systems

Since it was shown in 1971 [143] that the gauche preference characteristic of the anomeric effect can be reproduced by *ab initio* molecular orbital calculations on fluoromethanol, a large number of authors have carried out calculations, at various levels, on a range of simple molecules with the general formula XCH_2Y, where X and Y are electronegative groups, usually F, Cl, OH, OMe and NH_2. The preferred gauche, gauche (+*sc*, +*sc*) conformation of dimethoxymethane (see page 20), for example, is predicted by both *ab initio* and semi-empirical calculations [190–192] (though not by the extended Hückel and MINDO/2 methods [192]). In many cases the calculations reproduce all the main structural trends, including effects on bond length and angle which are discussed in detail below; and large scale calculations using extended basis sets give rise to structural parameters very close to those observed experimentally. It might be thought, therefore, that the electronic origin of the anomeric effect should be readily apparent, since it must be one of the elements built into the (fairly standard) calculations involved. This has turned out not to be the case [193], for various reasons: one of which is that it is possible to analyse the results in more than one way, depending in particular on whether the lone-pair electrons of oxygen are treated as energetically equivalent or non-equivalent. (Another, which must always be suspected in a situation of this sort, could be that no single electronic effect is so dominant that it alone can reasonably account for all the experimental observations.)

Table 11 shows how differences in the computed total energies of dihydroxymethane and various derivatives depend on the conformation about the two central C–O bonds. The data are all calculated by the same standard closed-shell self-consistent field method using the same (4–31G) basis set [190, 194–198], with standard bond lengths and angles. (Bond length optimisation changes the values for dihydroxy and dimethoxymethane very little [190]). Several trends are clearly apparent: the gauche *(sc)* conformation is generally a stable arrangement, except in (+*sc*, −*sc*) where it brings two groups into 1,3-diaxial type apposition: the relative stability of the (+*sc*, +*sc*) conformation decreases with increasing methyl substitution, as other, particularly steric, factors accumulate: and the preference for the *(sc)* over the *(ap)* arrangement is largest for the most electronegative element, fluorine, in $MeOCH_2F$. In the case of methoxymethyl formate the most stable conformation is (+*sc*, *ap*) rather than (+*sc*, +*sc*) and the next most stable is (+*sc*, *og*) (*103*) (*og* = orthogonal, indicating a dihedral angle of 90°), presumably as a consequence of the full sp^2-hybridisation at the ester oxygen.

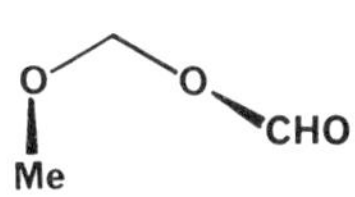

103

The conformations selected for Table 11 model various staggered conformations of axial and equatorial tetrahydropyran acetals, or α and β-glycosides, as shown in (*104–105*): the axial and equatorial conformations corresponding, respectively, to *synclinal (sc)* and *antiperiplanar (ap)* arrangements about the endocyclic C–O bond of the acetal group.

A complete description of the energy of the system as a function of the torsion angles RO–COR′ (Φ) and ROC–OR′ (θ) is given by a calculated potential energy surface (e.g. Fig. 10, for dimethoxymethane), and conveniently

Table 11. Calculated Relative Energies of $ROCH_2X$ as a Function of Conformation

Conformation		Relative Energy (Kcal mol^{-1})				
		R = R′ = R″ = H	R = Me, R′ = R″ = H	R = R″ = Me, R′ = H	R = R′ = Me, R″ = H	R = Me, R′ = CHO, R″ = H
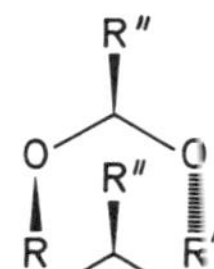	(+*sc*, +*sc*)	0	0	0	0	6.8
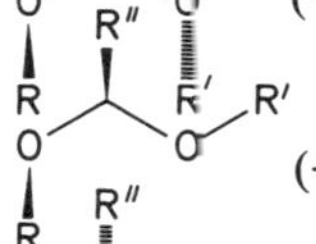	(+*sc*, *ap*)	4.4	4.3	4.83	2.59	0
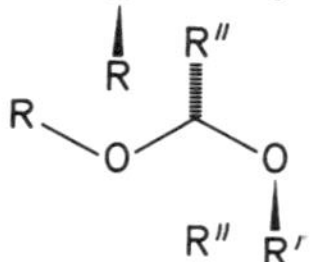	(*ap*, −*sc*)	4.4	3.0	3.40 (2.44)[a]	2.59	7.5
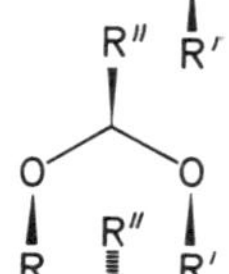	(+*sc*, −*sc*)	5.5	6.9	–	18.93	–
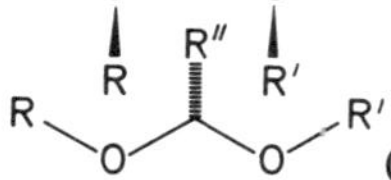	(*ap*, *ap*)	11.0	9.4	9.29	7.68	4.4
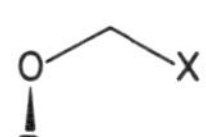			0 (X = F, Cl)			
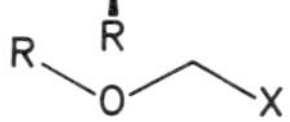			5.2 (X = F) 4.2 (X = Cl)			
(Ref.)		[194]	[195, 198]	[196]	[190]	[197]

[a] Figure in parentheses applies to the (*ap*, −*sc*) conformation, which is different from (*ap*, +*sc*) when R″ = Me

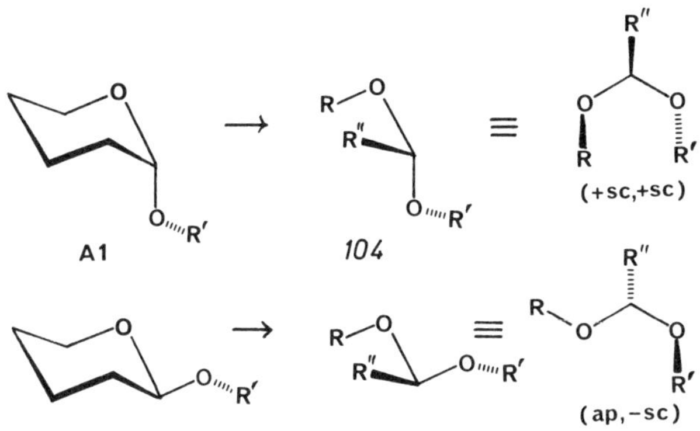

A1 104

E1 105

in two dimensions for fixed values of one torsion angle by sections through this surface. Figure 11 shows the relevant plot for dimethoxymethane with Φ fixed at 60° and 180°, corresponding to the axial and equatorial isomers of a tetrahydropyran acetal. The data in Table 11 correspond to minima on these plots. (For further discussion of Figs. 10 and 11 see below (page 67).

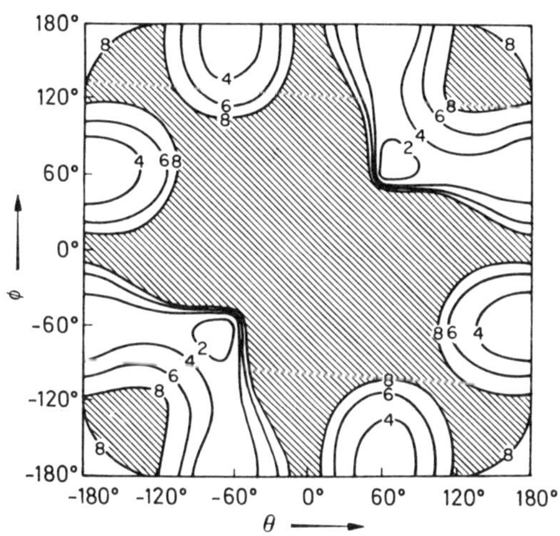

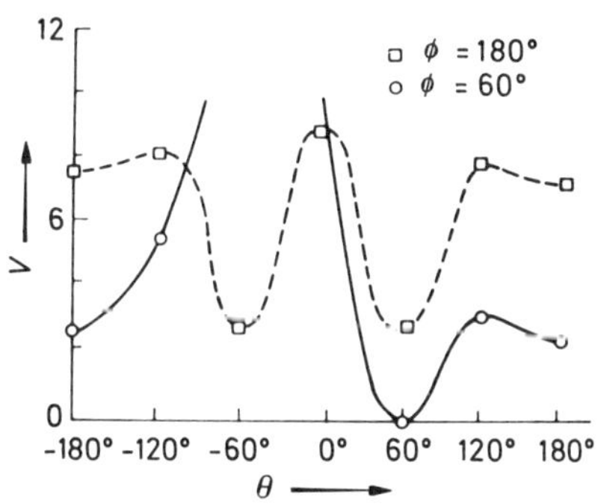

Fig. 10. Calculated potential energy surface for dimethoxymethane as a function of torsion angles ϑ and φ (see text). (Reproduced with permission from G.A. Jeffrey et al., J. Am. Chem. Soc. *100,* 373 (1978).) Copyright (1978) American Chemical Society

Fig. 11. Section through the potential energy surface of Fig. 10, showing the potential energy of dimethoxymethane as a function of torsion angle ϑ for fixed values of φ. (Reproduced with permission from G.A. Jeffrey et al., J. Am. Chem. Soc. *100,* 373 (1978).) Copyright (1978) American Chemical Society

It is important to recognise that these energy calculations refer to isolated molecules in the gas phase, whereas conformational equilibria are generally measured in the liquid phase, where medium effects can be substantial. Furthermore, calculated energy differences of the type shown in Table 11 represent relatively tiny differences between very large total energies. For methoxymethyl chloride, for example, the computed energies of the *sc* and *ap* conformations are −612.244362 and −612.237685 hartrees, the difference corresponding to 4.18 kcal mol^{-1}. In the circumstances the calculated differences are remarkably close to those observed. (For methoxymethyl chloride the anomeric effect is in the region of 3 kcal mol^{-1}, and certainly less than 4.2 kcal mol^{-1}, which is the measured height of the barrier for interconversion of the +*sc* and −*sc* conformations [198a]).

4. Effects on Bond Lengths and Angles

Although the clearest manifestation of the anomeric effect is the preference for gauche conformations about RO−CX bonds, detailed examination of the geometries of such systems shows that there are also characteristic patterns of bond lengths and angles associated with particular conformations. These patterns provide valuable evidence about bonding at the acetal and related centres, and, like the conformational preferences, are reproduced by *ab initio* and semi-empirical molecular orbital calculations.

It was shown many years ago that the carbon-halogen bonds in polyhalogenated methanes are shorter and stronger than those in the corresponding methyl halides [199]. The effect on the C−F and C−Cl bond lengths in the fluoro- and chloromethanes is shown by the data in Table 12; and is evidently cumulative, and greater for the C−F bond, even in absolute terms.

Table 12. Bond Lengths in Polyhalogenomethanes (data from Ref. [199, 200])

	C−X Bond Length (Å) in			
	CH_3X	CH_2X_2	CHX_3	CX_4
X = F	1.385	1.358	1.326	1.317
X = Cl	1.781	1.772	1.761	1.755

This shortening appears to be general for compounds with two or more electronegative substituents bound to the same carbon (or other) centre, as long as all the substituents are the same. It is apparent, for example, in the structure of dimethoxymethane, where the electron diffraction results are best interpreted in terms of substantially different bond lengths, 1.432 and 1.382 Å respectively, for the outer and inner C−O bonds [201]: and in the structures of glycosides, in which the majority of the C−O bond lengths at the anomeric (acetal) centre are shorter than the standard values of 1.431 and 1.416Å observed in methanol [202] and dimethyl ether [203], respectively (see Table 18, page 58).

When the electronegative substituents are different the pattern changes. Halomethyl ethers, $ROCH_2X$, have short $O-CH_2$ bonds but rather long C−X bonds [204]. However, measurements on cyclic systems show that this applies only to compounds with the favoured gauche conformation about the RO−CX bond [96]. The crystal structures of the *cis* and *trans*-2,3-dichloro-1,4-dioxanes (*106c* and *106t*) demonstrate the point nicely.

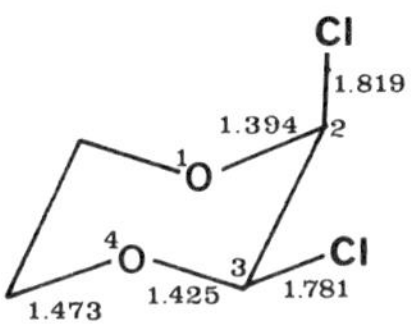

106c

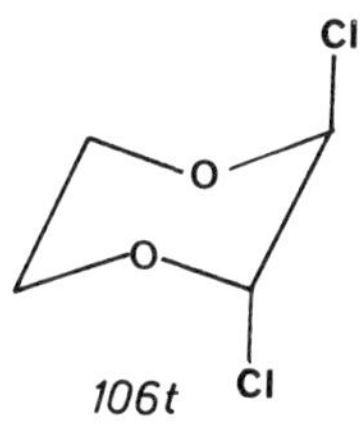

106t

In (*106t*) both C−Cl bonds to axial chlorine are long (1.833, 1.844Å), and the adjacent C−O bonds short (1.38Å mean value). The values are similar around C(2) of the *cis* isomer (*106c*), but in this compound the second chlorine atom has to be equatorial. As a result the O−C(3) and C(3)−Cl bond lengths (*106c*) revert to values which are normal for alkyl ethers and chlorides [96].

These patterns of behaviour are reproduced in detail in molecular orbital calculations. Most of the available results are summarised in Tables 13–16, which refer to calculations involving bond length and/or angle optimisation, usually at fixed values of the dihedral angles about the central carbon atom. (Where only selected bond lengths have been optimised the standard values

Table 13. Calculated Bond Lengths and Angles for Dihydroxymethane

Conformation[a]	*Bond Lengths* (Å) and		*Angles*	Ref.
	$HO-CH_2-OH$		OĈO	
(+*sc*, +*sc*)	1.421	1.421	–	[194]
	1.427	1.427	112.2°	[205]
(+*sc*, *ap*)	1.396	1.428	–	[194]
	1.408	1.437	(109° 28′)	[206]
	1.417	1.438	107.5°	[205]
(+*sc*, −*sc*)	1.416	1.416		[194]
	1.429	1.429	114.4°	[205]
(*ap*, *ap*)	1.409	1.409	–	[194]
	1.425	1.425	104.5°	[205]

[a] For details of conformations see Table 11 (R = R′ = R″ = H)

used for the others are given in parentheses.) The calculations clearly reproduce the bond shortening at the acetal centre (compare the lengths of O−CH_2 and CH_3−O bonds in Tables 14 and 15); and show also the larger shortening effect on the inner C−O bond associated with the lengthening of C−X in molecules CH_3OCH_2X, where X is either a more electronegative group (OAc, F) or from a higher row of the periodic table than oxygen (X = Cl) (Table 16).

Table 14. Calculated Bond Lengths for Methoxymethanol [195] and 1-Methoxyethanol [196]

Conformation[a]	*Bond Lengths (Å)*					
	Me−O−CH_2−OH			Me−O−CHMe−OH		
(+*sc*, +*sc*)	1.444	1.421	1.417	1.452	1.454	1.417
(+*sc*, *ap*)	1.444	1.400	1.430	–	–	–
(*ap*, +*sc*)	1.435	1.424	1.398	1.444	1.455	1.399

[a] For conformational drawings see Table 11, R = Me, R′ = R″ = H and R = R″ = Me, R′ = H

Table 15. Observed and Calculated Bond Lengths and Angles for Dimethoxymethane

Conformation[a]	*Bond Lengths (Å) and Angles*							*Ref.*
	Me−O−CH_2−O−Me				Ô	OĈO	$\hat{O}_2$	
(+*sc*, +*sc*)	*1.432*[b]	*1.382*[b]	*1.382*	*1.432*	*114.6*[b]	*114.3*[b]	*114.6°*	*(Expt.)*
	1.444	1.423	1.423	1.444	115.9°	113.9°	115.9°	[190]
	(1.43)	(1.40)[c]			(114)[c]	114.7°		[207, 208]
		1.376	1.376			114.9°		[209]
(+*sc*, *ap*)	1.443	1.400	1.425	1.434	115.8°	110.9°	116.1°	[190]
	(1.43)	(1.40)[c]			(114)[c]	109.3°		[207, 208]
		1.377	1.378			110.0°		[209]
(*ap*, *ap*)	1.435	1.406	1.406	1.435				[190]
	(1.43)	(1.40)[c]			(114)[c]	103.7°		[207, 208]
		1.378	1.378			104.6°		[209]

[a] For conformational drawings see Table 11 (R = R′ = Me, R″ = H)
[b] Experimental values, from electron diffraction measurements [201].
[c] Only angle OCO was optimised in this calculation.

Particularly interesting are the effects on bond lengths and angles of rotation about the C−O bonds, at the acetal centre. In the unsymmetrical (*sc*, *ap*) or (*ap*, *sc*) conformations the calculated bond lengths are substantially

Table 16. Observed and Calculated Bond Lengths for Methoxymethyl Halides and Esters

Conformation[a]	*Bond Lengths* (Å)						*Ref.*
	CH_3-O-CH_2-F			CH_3-O-CH_2-Cl			
(*sc*)	*1.421*[b]	*1.362*[b]	*1.385*[b]	*1.424*[b]	*1.362*[b]	*1.822*[b]	(*Expt.*)
	1.447	1.396	1.395	1.449	1.405	1.886	[198]
(*ap*)	1.437	1.405	1.376	1.439	1.410	1.837	[198]
	CH_3-O-CH_2-OCHO			$CH_3-O-CH_2-OCOCH_3$			
(+*sc*, +*sc*)	1.448	1.408	1.471	1.447	1.409	1.472	[197]
(*ap*, +*sc*)	1.434	1.413	1.441	1.434	1.413	1.441	[197]
(+*sc*, *ap*)	(1.43)	1.395	1.465	–	–	–	[197]
(*ap*, *ap*)	(1.43)	1.405	1.442	–	–	–	[197]

[a] Conformation about $MeO-CH_2X$ bond (see Table 11)
[b] Best experimental values from microwave measurements, ±0.006–0.008. Bond angles at C, O are 111.3, 113.5 and 112.9 and 114.0° (±0.50) for the fluoride and chloride, respectively (204).

different; Always the bond with the (*sc*) conformation is shortened compared with its length in molecules with *synclinal* conformations about both bonds; while the *ap* bond appears longer than in molecules with the *ap*, *ap* conformation (Tables 13–15). The bond angle at the acetal centre also varies significantly with conformation, falling in dimethoxymethane from about 114° (for +*sc*, +*sc*) to 110° for (+*sc*, *ap*) to 104° for the (*ap*, *ap*) conformation [207, 208]. A simple steric explanation of this last observation has been suggested by Gorenstein and Kar [207] who pointed out that non-bonded gauche interactions in the *synclinal* (*107*, Me:H+Me:OMe) and in the *anti-*

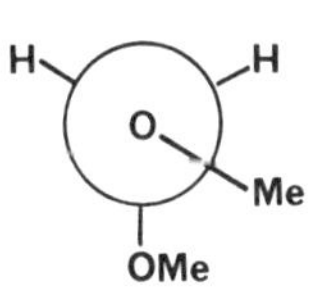

107

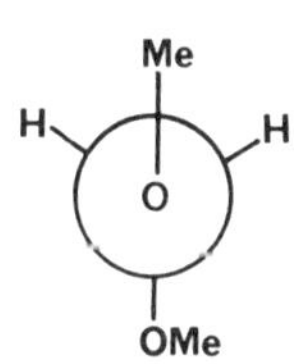

108

periplanar (*108*, 2 × Me:H) conformation, can be relieved by increasing (*107*) or decreasing (*108*) the central OCH_2O angle. If this explanation is correct the effect should be reduced when the gauche interactions involve hydrogen atoms rather than methyl groups, and there is some evidence in Table 13, and in experimental data given below (Table 18) that this is the case. On the other hand, the range of observed values for a wide range of compounds is so narrow that it seems likely that this effect only reinforces a basic stereoelectronic preference.

Calculations on the anion of dimethyl phosphate (summarised in Table 17)

Table 17. Calculated Bond Lengths and Angles for the Anion of Dimethyl Phosphate [207-8]

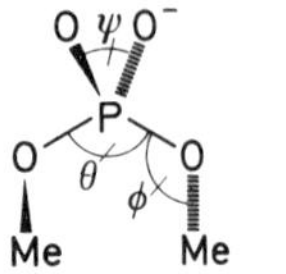	*Bond Lengths* (Å) and *Angles*						
	Me — O_1 — P — O_2 — Me				θ	φ	ψ
= (+*sc*, +*sc*)	(1.43)	1.667	1.667	(1.43)	98.81	125.7	112.4
(+*sc*, *ap*)[a]	(1.43)	1.666	1.667	(1.43)	94.90	124.3	111.8
(*ap*, *ap*)[a]	(1.43)	1.664	1.664	(1.43)	91.04	123.1	110.7

[a] Conformations as defined for dimethoxymethane (see Table 11)

make it clear that these effects are not confined to carbon centres. In particular the coupling between rotation about the RO−X bond and the RO−X−OR angle is of the same order of magnitude in the phosphate as in dimethoxymethane, and must be expected to be a general property of molecules RO−X−OR [207, 208].

The calculated bond lengths and angles shown in Tables 13–16 agree rather well with experiment, though comparisons are not entirely straightforward. The calculations refer to isolated molecules, so the closest comparison should be with data collected in the gas phase. There are often more uncertainties associated with such data than with those obtained from X-ray or neutron diffraction, but the electron diffraction data for dimethoxymethane (Table 15) and the microwave results for methoxymethyl fluoride and chloride (Table 16) show that the trends, if not the absolute values of bond length and angles are accurately reproduced.

By far the largest body of structural information concerning the detailed geometry of the acetal and related groups is to be found in the large number (over 500) of structures of carbohydrates determined by X-ray and neutron diffraction. Jeffrey and his group in particular [190, 194–198, 210] have carried out both structural investigations and calculations on models for the acetal centre, and agreement is remarkably good considering that the experimental values refer to molecules in the crystalline solid, while the calculations refer to the isolated molecule. The comparisons are shown in Table 18. As discussed above (page 49), the α-glycopyranosides have the (+ *sc*) conformation (*109*) about the O(1)−C(2) bond, while the β-anomers are *ap* (*110*): and the conformation about the exocyclic C(2)−O bond is generally gauche. So the

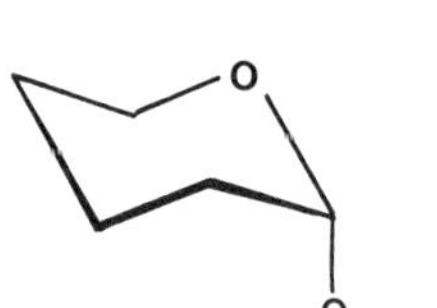

109

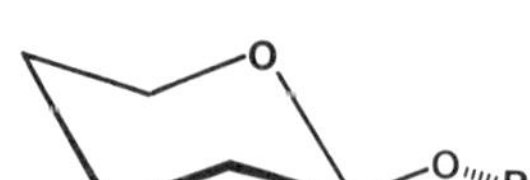

110

appropriate models are those with (+*sc*, +*sc*) and (*ap*, +*sc*) conformations, respectively.

The (*ap*, *ap*) conformation does not occur naturally, so the calculations cannot be tested for this case. But for the (+*sc*, +*sc*) and (*ap*, +*sc*) conformations agreement with theory is remarkably good. For pyranoses all but one of the hemiacetal group bond lengths is predicted accurately, to within one standard deviation, by the calculation for methoxymethanol, even though bond angles were not optimised. For methyl glycosides bond lengths are within 0.02Å of those predicted by the calculations (Table 18), though the observed OCO angles are 2–4° smaller (no doubt an effect of alkyl substitution, as discussed below for acetal *111*). The calculated difference in the lengths of the two C–O bonds at the acetal centre in the (*ap*, *sc*) conformation is confirmed most convincingly in the otherwise highly symmetrical tricyclic acetal (*111*) [213]. This molecule consists of three six-membered rings, one cyclohexane and two tetrahydropyrans, fused along a common bond and all in the chair conformation. Of five pairs of chemically equivalent bonds only the C–O bonds at the acetal centre differ significantly in length. This is clearly because the conformations about these C–O bonds are different: O(14) is axial while O(5) is equatorial to the other tetrahydropyran ring, so that the conformation about the acetal centre is (+*sc*, *ap*) in the direction C(4)–O(5)–C(6)–O(14)–C(13). The observed bond lengths and angles (Table 19) agree with those calculated for (+*sc*, *ap*) dimethoxymethane quite closely, (with the exception of the angle at the acetal centre, which shows the sharp decrease characteristic of *gem*-dialkyl substitution which is the basis of the Thorpe-Ingold effect on some cyclisation reactions [214]). And in particular the difference in the two C–O bond lengths at the acetal centre, resulting solely from their conformational inequivalence, is exactly that predicted.

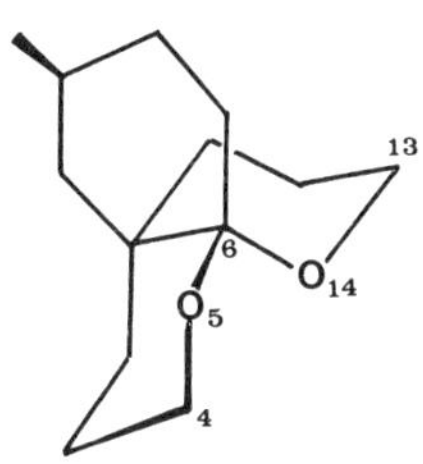

111

The largest discrepancies in bond lengths apparent in Table 18 are between observed and calculated values for 1-O-acetyl pyranosides, where the anomeric C–OAc bond is significantly shorter than calculated. It seems clear that this is because the ring oxygen of a glycopyranoside is effectively more electronegative than that of a simple tetrahydropyranyl or methoxy-group, presumably because of σ-electron-withdrawal (inductive effect) by the three or four hydroxy or acetoxy substituents on the ring. The effect is shown very clearly by the plot of Fig. 12, which compares the lengths of the exocyclic C–O bond of the acetal group in a series of α-glucopyranosides (*112*,

Table 18. Acetal Geometries in Pyranose Derivatives Compared with Bond Lengths and Angles Calculated for Simple Models

	Bond Lengths (Å) and				*Angles*[a]		
	C(5)–	O(1)–	C(2)–	OH	Ô(1)	OĈO	Ô(2)
α-D-Glycopyranoses[b]							
(*109*, R = H)	*1.439 (13)*	*1.427 (12)*	*1.400 (10)*		*113.7 (8)*	*111.6 (7)*	
Calculated[c,d]	1.444	1.421	1.417		–	–	
	1.452	1.454	1.417		–	–	
β-D-Glycopyranosides[e]							
(*110*, R = H)	*1.434 (7)*	*1.427 (8)*	*1.387 (12)*		*112.4 (9)*	*107.3 (7)*	
Calculated[c,d]	1.435	1.424	1.398		–	–	
	1.444	1.455	1.399		–	–	
	C(5)–	O(1)–	C(2)–	O–Me			
Methyl							
α-D-pyranosides[f]	*1.435 (9)*	*1.416 (3)*	*1.403 (7)*	*1.431 (6)*	*113.4 (9)*	*112.2 (5)*	*113.2 (2)*
Calculated[g]	1.444	1.423	1.423	1.444	115.9	113.9	115.9
Methyl							
β-D-pyranosides[h]	*1.433 (6)*	*1.428 (6)*	*1.383 (7)*	*1.427 (4)*	*111.4 (8)*	*107.4 (4)*	*113.4 (7)*
Calculated[i]	1.434	1.425	1.400	1.443	116.1	110.9	115.8
	C(5)–	O(1)–	C(2)–	O–Ac			
1-O-Acetyl-							
α-D-pyranosides[j]	*1.438 (4)*	*1.397 (7)*	*1.436 (9)*	*1.357 (15)*	*114.1 (10)*	*110.5 (13)*	
Calculated[k]	1.448	1.408	1.471	(1.36)	–	–	
1-O-Acetyl							
β-D-pyranosides[m]	*1.425 (4)*	*1.414 (6)*	*1.411 (5)*	*1.355 (10)*	*110.7 (3)*	*106.2 (5)*	
Calculated[n]	1.434	1.413	1.441	(1.36)	–	–	
	C(5)–	O(1)–	C(2)–	F			
α-D-Pyranosyl fluoride[o]	*1.453*	*1.365*	*1.386*				
Calculated[p]	1.447	1.396	1.395				
β-D-Pyranosyl fluoride[q]	*1.428*	*1.406*	*1.367*				
Calculated[r]	1.437	1.405	1.376				

	Bond Lengths (Å) and				*Angles*[a]		
	C(5)–	O(1)–	C(2)–	OH	Ô(1)	OĈO	Ĉ(2)
α-D-Pyranosyl chlorides[s,t]	*1.459*	*1.415*	*1.824*				
	1.443	*1.366*	*1.856*				
Calculated[u]	1.449	1.405	1.886				
β-D-Pyranosyl chloride[v]	*1.426*	*1.415*	*1.700*				
Calculated[w]	1.439	1.410	1.837				

[a] Figures in parentheses are standard deviations of means, in terms of last figure quoted.
[b] Mean values averaged over 28 structures for bond lengths [196] and 25 structures (bond angles) taken from the Cambridge Crystallographic Data File.
[c,d] Data from Table 14, for (+*sc*, +*sc*) (α) and (*ap*, +*sc*) (β) methoxymethanol and 1-methoxyethanol, respectively.
[e] Mean values for seven structures. Sources as in note b.
[f] Mean values for 8 structures [190]
[g] Data from Table 15, for (+*sc*, +*sc*) dimethoxymethane.
[h] Mean values for 9 compounds [190]
[i] Data from Table 15, for (*ap*, +*sc*) dimethoxymethane.
[j] Mean values of 9 compounds. Data from (197) and the Cambridge Crystallographic Data File.
[k] Data from Table 16 for (+*sc*, +*sc*) methoxymethylacetate.
[l] Bond length assumed.
[m] Mean value for 3 compounds. Sources as in note j.
[n] Data from Table 16 for (*ap*, +*sc*) methoxymethylacetate.
[o] Values for 2,3,4-tri-O-benzoyl-β-D-xylopyranosyl fluoride [42].
[p] Data from Table 16 for (*sc*) methoxymethyl fluoride.
[q] Values for 2,3,4-tri-O-acetyl-β-D-xylopyranosyl fluoride [42].
[r] Data from Table 16, for *ap*-methoxymethyl fluoride.
[s] Values for 2,3,4-tri-O-acetyl-6-deoxy-α-L-mannopyranosyl chloride [211].
[t] Values for 2,3,4,5-tetra-O-acetyl-α-D-mannopyranosyl chloride [211].
[u] Data from Table 16, for *sc*-methoxymethyl chloride.
[v] Values for 2,3,4-tri-O-acetyl-β-D-xylopyranosyl chloride [212].
[w] Data from Table 16, for *ap*-methoxymethyl chloride.

Table 19. Comparison of Observed and Calculated Geometries for Tricyclic Acetal (*111*)

	Bond Lengths (Å) and *Angles*						
	C(4)–O–C–O–C(13)				Ô(5)	OĈO	Ô(14)
Observed[a]	*1.432*	*1.412*	*1.438*	*1.432*	*115°*	*105.1°*	*115.5°*
Calculated[b]	1.443	1.400	1.425	1.434	115.8°	110.9°	116.1°

[a] Values taken from [213]
[b] Data from Table 15 [190]

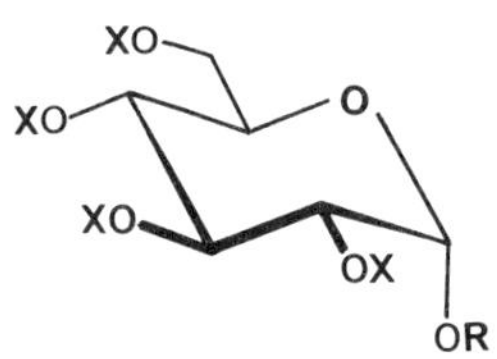

112

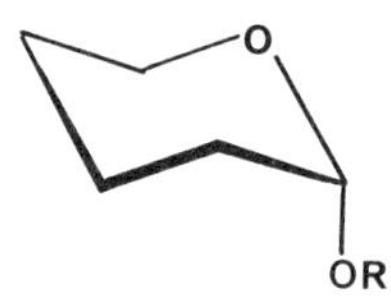

113

R = methyl, aryl, acetyl, X = H or acetyl) with that in a series of corresponding axial tetrahydropyranyl acetals, containing the structural unit (*113*, R = aryl or acyl). These bond lengths for the tetrahydropyranyl acetals show a linear correlation with the pKa of the conjugate acid (ROH) of the OR group [215]; and the calculated value for the corresponding bond of (+*sc*, +*sc*) methoxymethyl acetate [197] (shown by an asterisk on the plot of Fig. 12) is close to

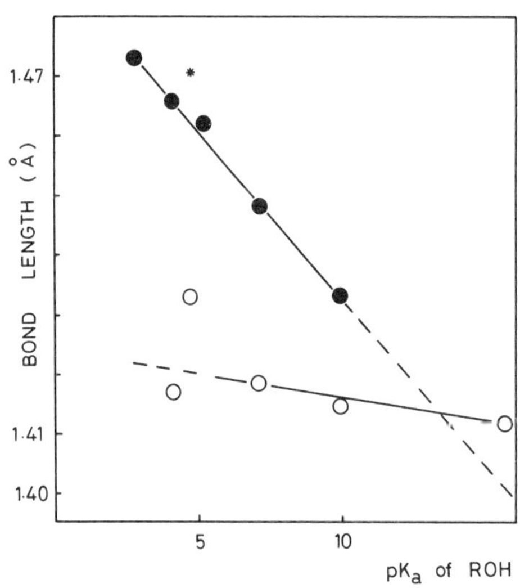

Fig. 12. Contrasting dependence of the length of the exocyclic C–O bond on the basicity of the axial OR group of α-glucopyranosides (open circles) and tetrahydropyranyl acetals (filled circles: see text). Data for phenyl [216] and p-nitrophenyl-α-D-glucoside [216, 217] and for α-glucose pentaacetate [216]: and for various aryl tetrahydropyran acetals [216–219]

that predicted by the plot for 2-acetoxytetrahydropyran. The correlation for the glucosides is less good, but the length of the glucosidic bond is clearly much less sensitive to the basicity of the OR group in this case (the slope of the line drawn is some 50 times smaller than that for the tetrahydropyranyl acetals). As a result the observed lengths of the C−OR bond in the two systems diverge strongly for more weakly basic (more electronegative) OR groups, until for the acetoxy derivatives (pKa of acetic acid 4.76) the difference is expected to be about 0.04Å, almost exactly the difference (0.038Å) between the bond length observed for α-glucose pentaacetate (C−OAc = 1.433 Å [216]) and that predicted for (+ *sc*, + *sc*) methoxymethyl acetate (Table 18).

Further implications of these and related observations are discussed below (page 90), but the idea that the ring oxygen of glycopyranosides is effectively more electronegative than that of simple alkoxy groups in acetals is consistent with several different lines of evidence. Of particular relevance to the discussion of this section is that it leads to the prediction that the C−O bond lengths at the acetal centres of methyl glycosides will be different even in the (+*sc*, +*sc*) conformation, where calculations on dimethoxymethane must predict equal bond lengths. In practice the endocyclic acetal C−O bond is invariably longer than the corresponding exocyclic bond in methyl pyranosides and in pyranoses: and, as shown in Table 18, the exocyclic bond is significantly shorter than calculated in every case.

5. Effects in Furanoses

Although the anomeric effect is expected to operate through the acetal centre of furanoses as well as pyranoses, the conformational consequences are much less dramatic, and as a result relatively little work has been done in this area. In principle the bond to an electronegative substituent (X) at the anomeric centre of a furanose (e.g. *114*) stands in the same relationship to the lone

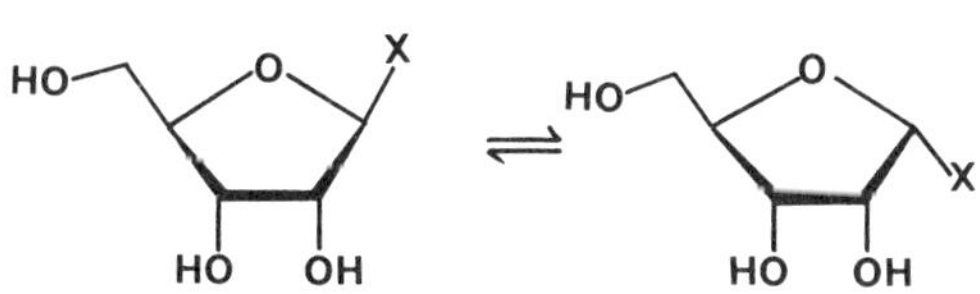

114

pair electrons on the ring oxygen whether it is in the α or β-position. In practice the puckering of the ring usually [221] adjusts so that the substituent is quasi-axial in either case [9c]. The conformation about the exocyclic C−O bond, on the other hand, is subject to the exo-anomeric effect in the usual way [67].

The most extensive structural studies on furanosides involve nucleosides and nucleotides (e.g. *115*), where there is good evidence for conformational preferences and patterns of bond lengths arising from the anomeric effects [222–224]. The C(1)−N(1) bond length for a series of nucleosides, for example,

Table 20. Bond Length Changes at the Anomeric Centre of 5-Substituted Uridines, as a Function of the Substituent [222].

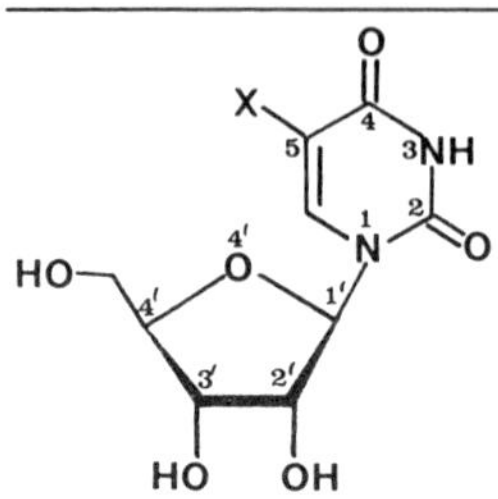

115

Substituent, X	*Bond Length* (Å)		
	C(4')– O(4')–	C(1')–	N(1)
NH_2	1.460	1.432	1.427
Me_2N	1.465	1.422	1.464
HO	1.460	1.416	1.467
MeO	1.463	1.395	1.490
Me	1.460	1.410	1.481
H	1.454	1.413	1.490
Cl	1.448	1.410	1.474
NO_2	1.461	1.390	1.504

is related to the dihedral angle C(4)'–O(4)'–C(1)'–N(1), and also to the electronegativity of the heterocyclic ring, as transmitted through N(1). Thus there is clear evidence [224] that the length of the C(1)'–N(1) bond of 5-substituted uridines (*115*) is increased, and that of O(4)'–C(1)' decreased, in compounds where X is electron withdrawing (Table 20). This behaviour is similar in pattern and magnitude, and presumably also in origin, to the bond length changes observed in 2-aryloxy tetrahydropyrans discussed above (page 60).

6. The Stereoelectronic Basis of the Anomeric Effect

Since both the conformational preferences and the pattern of bond lengths associated with the anomeric and exo-anomeric effects are reproduced in detail by the best *ab initio* calculations, there is no doubt – notwithstanding the discussion on page 49 that if we can find the right way to analyse the results, the electronic basis of the effect can be explained. Furthermore, since it is clear that both the geometries and the energies of lone pair-orbital interactions are likely to be involved, we know that the analysis should be carried out in terms of canonical molecular orbitals (see page 41).

We start by considering the electronic structure of FCH_2F, a useful model which illustrates the relevant lone pair-orbital interactions, but has the advantage that the important non-bonding electron pairs on fluorine are equivalent. The consequences of the rotational asymmetry about the C–O bonds of $RO-CH_2X$ and subsequently $RO-CH_2-OR$ can then be understood as restrictions on the bonding pattern established for FCH_2F.

a. The Structure of CH_2F_2

The relevant molecular orbitals of methylene fluoride are illustrated in Fig. 13. (These representations are strictly qualitative, and only selected orbitals with actual or potential π-symmetry are included.) Of the three p-orbitals of the central carbon atom two are principally engaged in σ-bonds with the two fluorine atoms, giving rise to two equivalent σ_{C-F} bonding orbitals, and the corresponding antibonding pair (σ^*_{C-F}). **(Only one of each is illustrated.)** The third, p_Z, orbital of carbon is involved in bonding the two hydrogen atoms of the CH_2 group, as in π_1 and π_3, but it also has the correct symmetry for π-type overlap with the p_Z-lone pairs of both fluorine atoms. The result is a significant π-contribution to bonding in the CF_2 group, which results in C−F bonds which are stronger and shorter than those in monoalkyl fluorides like CH_3−F.

This extra π-bonding may be seen simply as a result of the more extensive delocalisation possible in π_1 of CH_2F_2, compared with CH_3F: or more formally as a result of the mixing of π_4* with π_3 (and to a lesser extent with π_1) which allows the donation of electron density from the p_Z-orbitals of fluorine to the antibonding orbital of the methylene group. (An equivalent description in terms of the interaction of the fluorine and CH_2-group π and π*-orbitals has been given by Epiotis et al. [225], and similar descriptions in terms of a dissection into FCH_2 and F fragments by Eisenstein et al. [226] and Wolfe [193].) We will refer to orbital interactions of this type, in accordance with previous practice [193], as n-π* overlap.

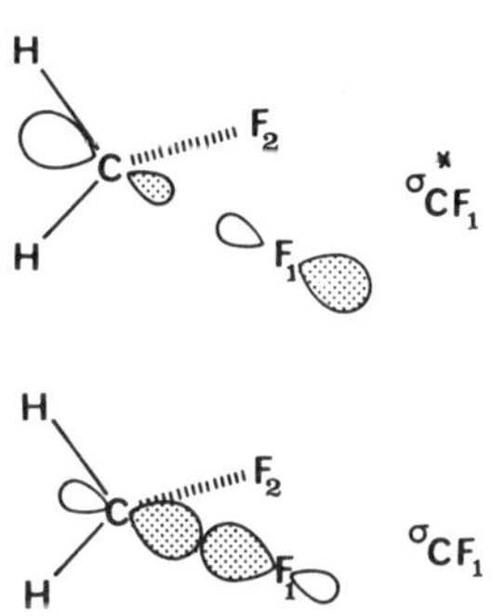

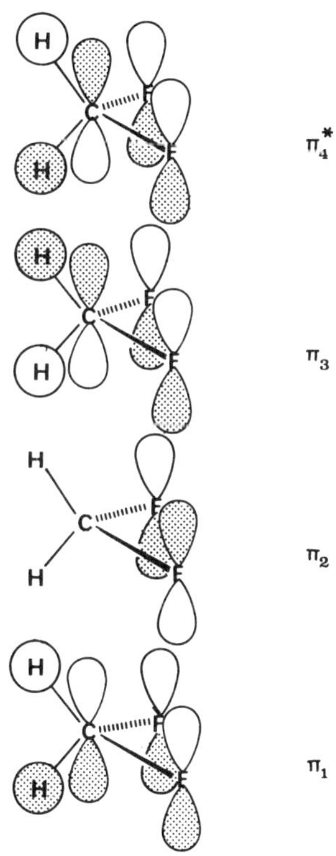

Fig. 13. The bonding orbitals of FCH_2F which involve the carbon $2p_x$, $2p_y$ and $2p_z$ orbitals

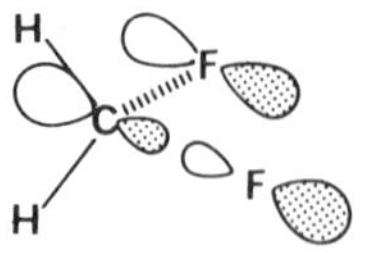

Fig. 14. The n_π-σ^* interaction in FCH_2F

Apart from this π-type overlap, between the p_Z orbitals of carbon and the two fluorine atoms, a second, independent type of extra overlap is possible, involving the fluorine p-type lone pairs lying in the FCF plane. These each have the correct symmetry (Fig. 14) for overlap with the σ-orbitals of the other C−F bond, and the two-electron interaction with the antibonding σ^*_{C-F} orbital is stabilising [225, 226].
Thus two separate and quite independent types of π-overlap (n-π^* and n−σ^*) contribute extra stabilisation to CH_2F_2, and thus to the observed strengthening of the C−F bonds compared with CH_3F. The n-π^* overlap shown in Fig. 13 clearly contributes to the observed shortening of the C−F bonds, since both gain some π-character. The n-σ^* overlap (Fig. 14) will also lead to a shorter bond to the donor atom, but populating the σ^* orbital will in principle weaken, and thus lengthen, the bond to the other atom. If the two bonds are equivalent, and the interaction is stabilising, presumably the result in this case also will be some shortening of both bonds. The most relevant calculations, by Wolfe and his co-workers [193], using a PMO analysis based on the fragmentation FCH_2...F, suggest that the n-σ^* interaction is larger, mainly because σ^*_{C-F} is lower in energy than the π^* orbital of the CH_2F fragment.

b. Bonding in $ROCH_2X$

Compared with fluorine, oxygen has only one, rather than two non-bonding pairs of p-type electrons, but this is of substantially higher energy than the lone pairs of fluorine, and may therefore be expected to be more effective in π-donor interactions with adjacent vacant orbitals of the correct symmetry. Conversely the antibonding σ^*_{C-O} orbital is of higher energy than σ^*_{C-F}, and thus, other things being equal, a less effective acceptor. The combination of these two properties makes α-halogenoethers quite different in reactivity from the very stable acetals and methylene halides (see below, page 80). Oxygen acts almost exclusively as a π-donor, and the carbon-halogen system almost

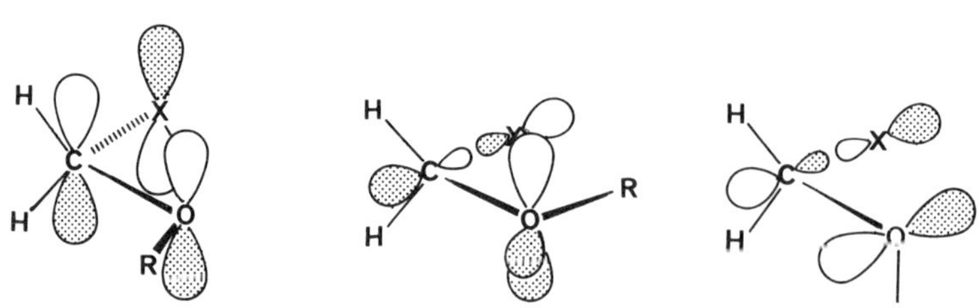

116 *117* *118*

Fig. 15. The geometry of the n_π-π^*, n_σ-σ^* and n_π-σ^* interactions in $ROCH_2X$

exclusively as an acceptor. But the efficiency and mechanism of the donor-acceptor process depend on the conformation around the RO-CX bond.

As shown in Fig. 15 n-π^* overlap, of the type illustrated for CH_2F_2 in Fig. 13, requires the R−O−C−X system to be planar. Of the two possible planar conformations, with dihedral angles of 0 and 180°, the latter (*116*) is likely to be preferred, because it avoids the eclipsing interaction between R and X.[5].

This means that π-donation by oxygen into the σ^*_{C-X} orbital requires a rotation of 90° about the C−O bond, to give (*118*). Now if this n-σ^*-type interaction is dominant in CH_2F_2, and since oxygen is both a better donor and a poorer (σ^*_{C-O}) acceptor than F, we clearly must expect the most important lone-pair interaction in $ROCH_2X$ to be n-σ^*_{C-X}, leading to a preference for conformations close to (*118*), rather than to (*116*).

In practice conformational equilibria are determined by a balance of several independent factors, and it is instructive to analyse calculated potential energy profiles for rotation about the C−O bonds of compounds $RO-CH_2X$ to try to understand how this balance is struck. A relatively simple technique which has been used for this purpose is a Fourier analysis of the total rotational potential energy function (V_T) in terms of three components V_1, V_2 and V_3, representing one-fold, twofold and threefold rotational barriers, respectively [114]. Internal rotation in ethane (or methanol), for example, is completely defined by V_3 alone, but less symmetrical molecules like $ROCH_2X$ require more functions: which, although the analysis is a formal one can sometimes be 'identified' with known physicochemical factors.

Calculated potential energy functions (V_T) describing rotation about the $RO-CH_2X$ bonds of fluoromethanol and methoxymethyl fluoride are shown in Figs. 16a and b respectively, together with their breakdown into the three components V_1, V_2 and V_3. The threefold barrier V_3 accounts for the usual preference for staggered ($\varphi = 60$, 180°) over eclipsed 0, 120°) conformations, and is substantially larger (maximum −3.90 kcal) in methoxymethyl fluoride than in fluoromethanol (−0.96) because the eclipsing interaction involves a methyl group rather than a bond to hydrogen. The onefold barrier indicates a preference for the (*sp*) over the (*ap*) conformation (*119*, R = H preferred to *120*, R = H). This probably results from stabilisation of (*119*, R = H) by

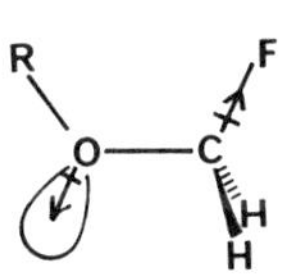

119

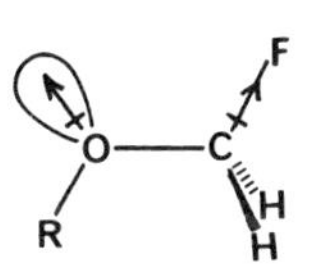

120

---

[5] This argument ignores the possible overlap of the σ-type lone pair on oxygen with σ^*_{C-X} in the eclipsed conformation (*117*). As usual we assume that the difference in energy between the σ and π-type lone pairs is so great that donor-acceptor interactions involving the latter are likely to be much more significant [158, 225].

O–H–F hydrogen bonding, not possible in (*120*), and is also consistent with the usual preference for the form with the smallest resultant dipole moment (the major contribution at 0 being due to the lone pair electrons, giving moments in the directions indicated in *119* and *120*). Note that this V_1 term, which is dominant in the simplest model, FCH_2OH [114], is much reduced in methoxymethyl fluoride [198] (from 5.25 to 1.61 kcal mol^{-1}) and it is actually reversed in methoxymethyl chloride [198], where the V_1 term favours the (*ap*) conformation, corresponding to (*120*).

Stereoelectronic effects involving n-π* and n-σ* interactions are expected to show up in the twofold component V_2, and this goes through a minimum.

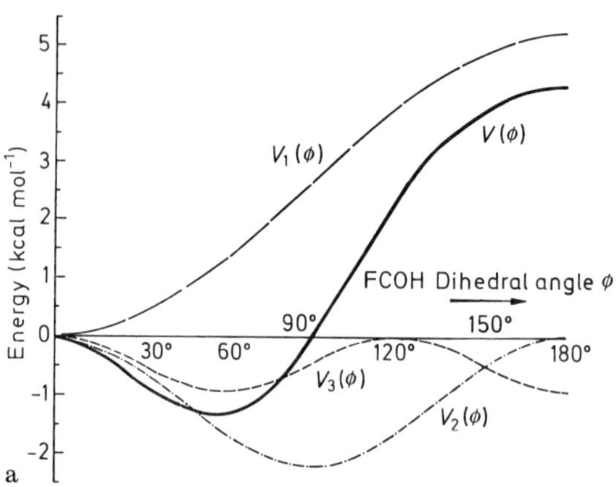

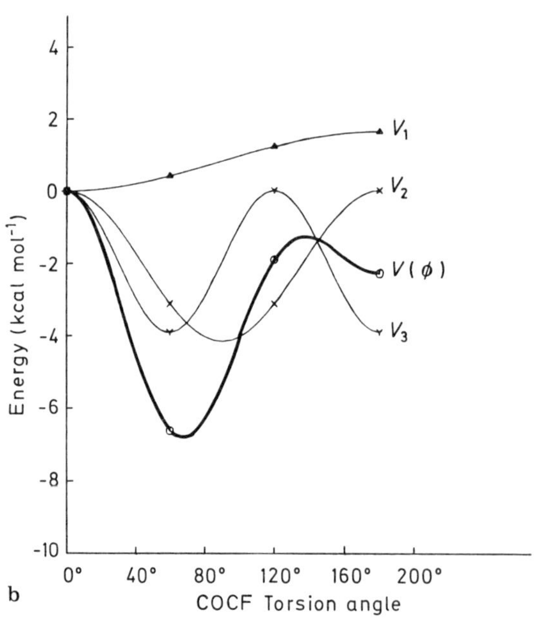

Fig. 16a and b. Fourier decompositions of the potential functions for (a) $HOCH_2F$ and (b) $MeOCH_2F$, as a function of torsion angle. $V(\varphi)$ is the total energy and $V_{1,2,3}$ (φ) are the one, two and three-fold rotational barriers (see text). (Reproduced with permission from **a** L. Radom, W.J. Hehre and J.A. Pople, J. Am. Chem. Soc. *94*, 2371 (1972); and **b** G.A. Jeffrey and J.H. Yates, J. Am. Chem. Soc. *101*, 820 (1979).) Copyright (1972, 1979) American Chemical Society

for the orthogonal (90°) conformation in both cases, which is almost twice as deep (−4.15 compared with −2.20 kcal mol^{-1}) for the methoxymethyl fluoride. This is consistent with greater stabilisation from n-σ^* (*118*) than from n-π^* overlap (*116*) as discussed above.

This analysis gives us some idea how these three familiar factors – steric, dipole-dipole and stereoelectronic interactions – contribute to the total potential energy function $V(\varphi) = V_1(\varphi) + V_2(\varphi) + V_3(\varphi)$. Two of the three component curves (V_1 and V_2) have energy maxima at $\varphi = 180°$, together with minima at $\varphi = 0$ and $\varphi = 90°$, respectively. The most stable conformation will therefore clearly be close to the $\varphi = 60°$ minimum (rather than the equivalent 180° minimum) of the V_3 function. The calculated position of minimum energy thus depends on the magnitudes of the coefficients V_1–V_3 in each case. For fluoromethanol [114] it is at $\varphi = 55°$, for methoxymethyl fluoride at $\varphi = 67°$ (64° with geometry optimisation) [198]. In practice, of course, all three factors will be affected differently by the medium, and a different balance may be struck. But the clear prediction from the calculations for molecules $RO{-}CH_2X$ is of a near-gauche (*sc*) conformation, as observed; with the usual preference for staggered geometry reinforced by dipole-dipole interactions and a stereoelectronic effect which is predominantly n-σ^* in origin. And the trends in the potential functions for the series $HOCH_2F$, $MeOCH_2F$ (and $MeOCH_2Cl$) suggest that the last factor is likely to be dominant in real molecules.

c. Bonding and Restricted Rotation in Dimethoxymethane

Because of its lower symmetry the potential energy function describing rotation about the inner C−O bonds of $MeO{-}CH_2{-}OMe$ is more complicated, and has to be presented in three dimensions (Fig. 10, above). Sections through the energy surface, corresponding to fixed conformations about one C−O bond (e.g. Fig. 11) give potential energy curves for rotation about the other, similar to the V_T curves shown in Fig. 16; but these cannot in general be analysed simply in terms of three components because of the lower symmetry of dimethoxymethane. For the Fourier analysis fourth and fifth terms are necessary, except where the dihedral angle is fixed at 180°. The relevant

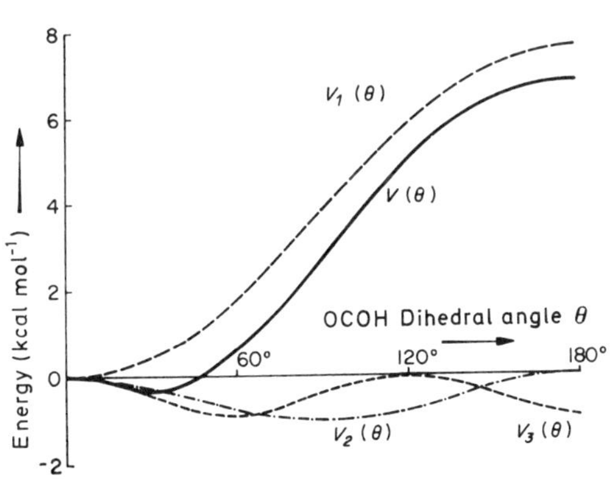

Fig. 17. Fourier decomposition of the potential function for dihydroxymethane, with one H−O−C−O torsional angle fixed at 180°. $V(\theta)$ is the total energy and $V_{1,2,3}$ (θ) are the one, two and three-fold rotational barriers (see text). (Reproduced with permission from G.A. Jeffrey, J.A. Pople and L. Radom, Carbohydrate Research *25*, 117 (1972).)

analysis of the rotational function under these conditions has been performed for this form of $HO-CH_2-OH$ [194], and is illustrated in Fig. 17.

The rotational function is dominated by the onefold component, V_1, which is a minimum at $\theta = 0°$ partly because the dipoles associated with the lone pair electron density are opposed in this (*ap, sp*) conformation (*121*) [194], but also no doubt because the weak hydrogen bonding possible in (*121*) but not the (*ap, ap*) conformation (*122*) is a significant source of stabilisation for an isolated molecule [205].

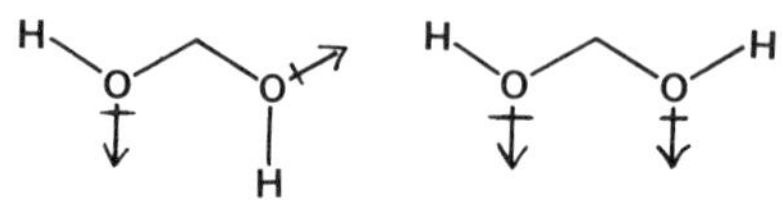

121 *122*

The analysis for dimethoxymethane [198], with the conformation fixed gauche about one acetal C–O bond, is illustrated in Fig. 18; which is included primarily to show that interpretation is no longer even relatively simple. Again the major contribution to the rotational barrier comes from the V_1 term, which is now a *maximum* at 0° (the (+*sc, sp*) conformation, *123*) and a minimum at 180° (+*sc, ap, 124*).

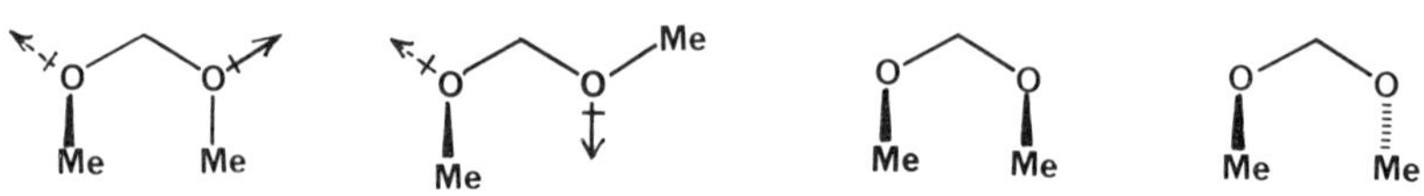

123 *124* *125* *126*

In this case the V_1 term obviously contains no contribution from H-bonding, but it is dominated rather (as probably is the V_4 term also) by the strong 1,3-diaxial-type interaction between the parallel O-Me bonds in conformation (+*sc*, −*sc*) (*125*), which is a maximum at $\theta = 60°$. (The dipole moment is a minimum in conformation (+*sc*, +*sc*) ($\theta = +60$, *126*) (for a discussion see de Hoog *et al.* [34]) and a maximum at $\theta = 120°$, so it too probably appears mainly in the V_4 term, and so some extent in V_5.)

Overall the smaller V_3 and more substantial V_2, V_4 and V_5 terms, which have minima at +60, 90, 90 and 45° respectively, moderate the effect of V_1 to give a resultant angle of minimum energy for rotation about the second bond of 76°, so that the predicted most stable conformation is close to (+*sc*, +*sc*) (*126*) as found experimentally (see above, page 00) [190, 198].

In dimethoxymethane, therefore, which is the first acetal to be even partially analysed in this way, steric effects override dipole-dipole interactions, but there remains a clear preference for near-gauche conformations (*ap, sc* and +*sc*, +*sc*) about any acetal C–O bond which is not fixed. And this preference, though the result of a number of terms, is determined principally by V_4 (singlefold) and the twofold components V_2 and V_5, which contain important contributions from n-σ^* stabilisation.

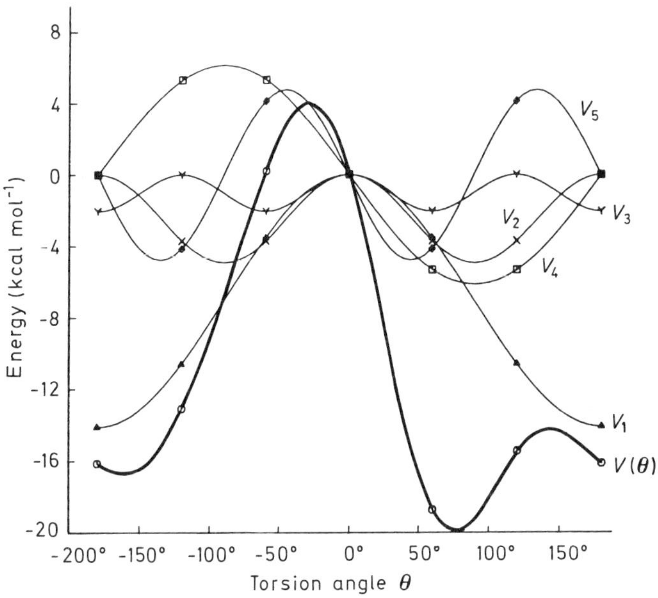

Fig. 18. The complex Fourier decomposition of the potential function for dimethoxymethane, with one Me–O–C–O torsion angle fixed at 60°. The symbols have the same meanings as in Fig. 17. $V_{4,5}$ are the extra (sine) terms needed – see text. (Reproduced with permission from G.A. Jeffrey and J.H. Yates, J. Am. Chem. Soc. *101*, 820 (1979).) Copyright (1979) American Chemical Society

The extent and nature of the π-contributions to bonding in the acetal group will depend on the conformation about the two inner C–O bonds much as illustrated for RO–CH_2X in Fig. 15 (above). In the (*ap*, *ap*) conformation (*127*) n-π^* overlap is optimal, but n-σ^* interactions are prevented. In the (*ap*, *og*) conformation (*128*) the *ap* bond has minimal π-character because the extended π-system of (*127*) has disappeared, but n-σ^* overlap is now possible across the other bond. Finally, where the dihedral angle is 90° about both the C–O bonds at the acetal centre (+*og*, +*og*, *129*), n-σ^* interactions are possible across both bonds, but n-π^* overlap is precluded.

This picture of bonding accounts satisfactorily for the calculated and observed patterns of bond lengths in the different conformations of acetals

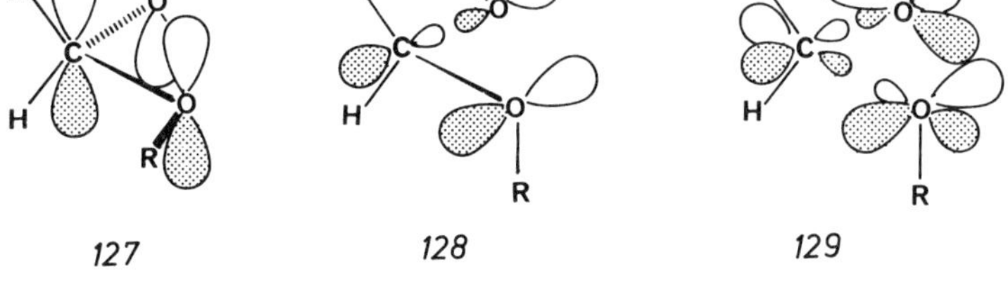

127 *128* *129*

(*130–132*)[6]. The bond shortening calculated (the conformation is not observed experimentally) for the (*ap*, *ap*) form (*130*) is explained by the n-π^*-type of π-bonding in the O−C−O system (*127*). The observed asymmetry in conformation *ap*, *sc* (*131*, close to *128*), which has a very short (*sc*) bond and the longest (*ap*) bond, is explained by the minimal π-contribution across the latter and n-σ^* overlap across the former: and the two-way n-σ^* overlap possible

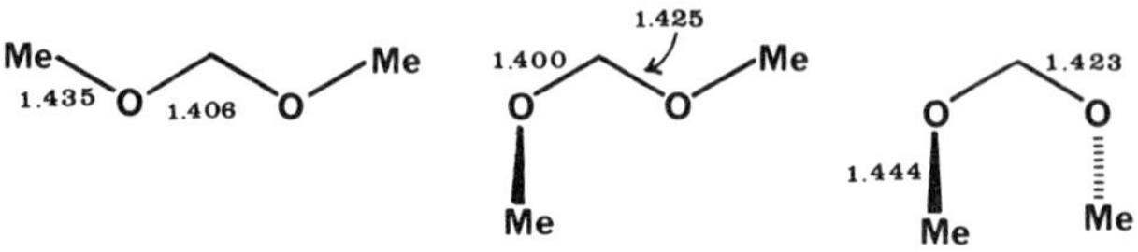

*130** *131** *132**

in the (+ *sc*, + *sc*) conformation (*132*, close to *129*) means that both C−O bonds at the acetal centre are subject to the bond lengthening effect of populating the σ-antibonding orbital: which largely offsets the bond shortening of the π-bonding contribution.

This is the picture for a symmetrical acetal. In an unsymmetrical structure $ROCH_2OR_X$ the more electronegative oxygen (OR_X) will be the poorer n-donor, but a better σ-acceptor (lower energy σ^* orbital). As a result bond shortening will be greater in the C−OR bond and bond lengthening greater in the C−OR_X bond, for conformations (*133*, *134*, corresponding to axial tetrahydropyranyl acetals, or α-glycosides) in which the conformation about the RO−C bond is gauche, and OR can act as a donor in the n-σ^* interaction. Conversely, in conformations (*135*) where only the more electronegative oxygen OR_X can act as a donor, the shortening of the C−OR_X bond and the lengthening of C−OR will be suppressed, as it is in equatorial tetrahydropyranyl acetals and aryl and acyl-β-glycosides (see below, page 90). The extreme cases, where OR_X is strongly electronegative, correspond closely to the alkoxymethyl halides, $ROCH_2X$, which have very short RO−CH_2 bonds in the (*sc*)-conformation and significantly longer ones in the (*ap*) form (Tables 16 and 18, above), where only n-π^* overlap is possible.

The bond angles at oxygen also reflect the degree of π-participation. Typical angles for a wide range of acetals are $\alpha \simeq 114°$ for acyclic and tetrahydropyranyl acetals, which adopt conformation (*133*), and $\beta \simeq 111°$ for the corresponding angle of an (*ap*) bond (*135*). And in the *bis*-acetal (*136*), in

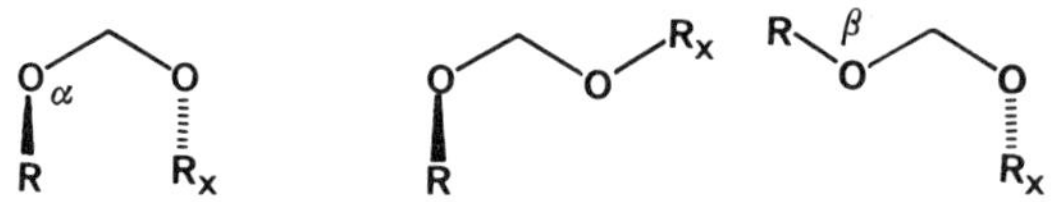

133 *134* *135*

[6] The bond lengths shown in (*130–132*) are calculated [190], and comparable with observed values (Tables 15 and 18, above).

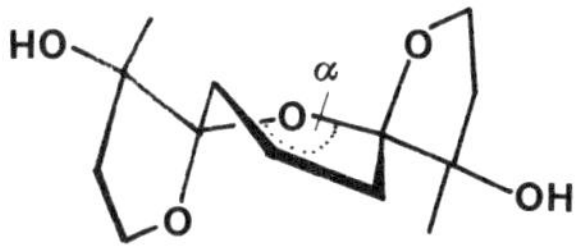

136

which the geometry allows the central oxygen atom to act as an n-donor to two adjacent C−O bonds the corresponding OCO angle α is 119°, as expected for a fully sp^2-hybridised centre [227].

d. The Generalised Anomeric Effect

Thus the preference for gauche conformations and the pattern of bond lengths and angles characteristic of the anomeric effect can be accounted for, at least qualitatively, in terms of lone pair-antibonding orbital interactions, of which the n-σ^* variety is clearly the more important. The relative magnitudes of the related effects in systems other than acetals can therefore be predicted from the relative energies of the relevant lone pair and antibonding orbitals. Some calculations of Wolfe, Whangbo and Mitchell [193] show that the strengths of n-π^* interactions depend very little on the nature of the C−X bond in (*ap*) $HOCH_2X$, but that n-σ^* stabilisation (in the *og* conformation) does depend on X, in the order $Cl > F \gg CH_3$. This is the ascending order of energies of the antibonding orbitals, $\sigma^*_{C-Cl} < \sigma^*_{C-F} < \sigma^*_{C-C}$: (the levels of σ^*_{C-X} generally decrease with increasing electronegativity of X in a given row of the periodic table, and with increasing atomic number within a group). This is consistent with the relative magnitudes of the anomeric effect in all the sequences given in Section A, with the exception of the position of the R_3N^+ substituent, which is discussed in the following Section. The observation that the anomeric effect is comparable in 2-alkoxytetrahydropyrans and the corresponding thia-compounds (page 00, above), although the n_π-orbital of S is undoubtedly of higher energy, is readily explained [193] by the less efficient overlap of the sulphur 3p orbital with the poorly-matched 2pσ^* orbital.

e. The Reverse Anomeric Effect

The reverse anomeric effect as originally definea (page 17) refers to the preference of R_3N^+ groups at the anomeric centre of pyranosides of the equatorial position. There is only a little experimental evidence about the effect, and no detailed structural information, so a very full discussion is not justified. But it is clear that the effect is not predicted by our analysis of the stereoelectronic basis of the anomeric effect in systems $ROCH_2X$, which would predict that R_3N^+, having no lone pairs but a σ^*_{C-N} bonding orbital slightly lower in energy than σ^*_{C-O}, should act exclusively as an electron-acceptor, favouring a gauche orientation about the RO−CH_2 bond. Nor is the effect predicted by molecular orbital calculations. No relevant computations are available for $HOCH_2NH_3^+$, but recent work [228] on $(HO)_2CHNH_3^+$ is consistent with our general conclusions about the nature of the anomeric effect.

The simplest explanation is that the anomeric effect is reversed because dipole-dipole interactions are different, and no longer reinforce the stereoelectronic preference. In the gauche conformation (*137*) of $ROCH_2X$, where the lone pairs on X are not specifically directional, the dipole moment associated with the lone pair electron density on oxygen opposes that of the C−X

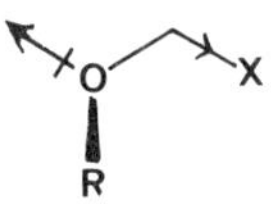

137

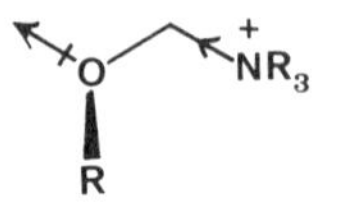

138

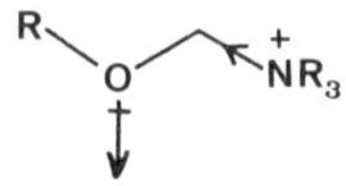

139

bond, and the resultant dipole moment is a minimum. In the corresponding conformation of $ROCH_2NR_3^+$ (*138*) the C−X dipole is reversed, and the resultant moment is now a minimum in the *antiperiplanar* conformation (*139*). This explanation may suggest that our analysis of the anomeric effect underestimated the contribution of dipole-dipole interactions, but the argument used here is specifically relevant only to very unsymmetrical systems $ROCH_2X$ (in acetals the dipole moments of the central C−O bonds cancel each other out): and the known instances of equatorial preference of R_3N^+ groups are all cases where there are powerful steric reasons for this preference (see page 16).

Finally, we should mention the equatorial preference of neutral nitrogen substituents like NH_2 (see above, page 15), which is sometimes referred to as a reverse anomeric effect [193]. These too are very unsymmetrical systems, but now in the reverse sense, since in the $RO-CH_2-NH_2$ group the nitrogen lone pair is the best donor and σ^*_{C-O} is the best acceptor. As a result the conformational consequences will principally concern rotation about the C−N bond, and the gauche preference about $RO-CH_2$ will be relatively small. And, in particular, not large enough to overcome the usual steric preference for *antiperiplanar* (or equatorial) conformations.

f. The Exo-Anomeric Effect

The exo-anomeric effect (see above, page 18) is the preference for the gauche (*sc*) conformation about the glycosidic C−OR bond of sugar derivatives, and is in principle identical to the anomeric effect, which is defined in terms of the preference for the same conformation about the inner (ring) C−O bond of the acetal group.

7. Related Stereoelectronic Effects

a. The Anomeric Effect at sp²-centres

The overlap between non-bonding electron pairs on oxygen and the antibonding σ^*-orbitals of adjacent bonds, which is the basis of the anomeric effect, may also account for the well-defined conformational preferences often observed in planar systems based on sp^2-hybridised carbon. It is well-known, for example, that esters and acids show a clear preference for the Z (*140*) over the E (*141*) conformation [229–231], even though the latter would be expected to be favoured on steric grounds, at least for formate esters.

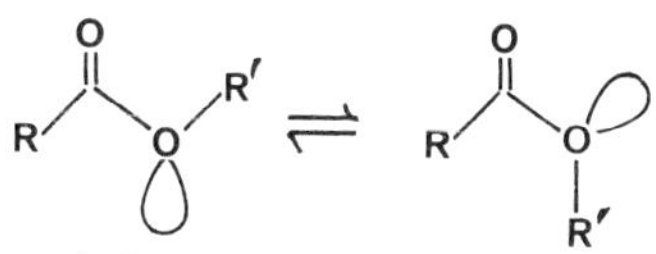

140 141

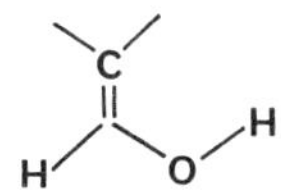

142

This preference, which is reproduced in suitable molecular orbital calculations [156, 232, 233], is likely to be a stereoelectronic effect related to the anomeric effect [234, 235], since only the favoured Z-conformation has a non-bonding electron pair *antiperiplanar* to the C–O bond. (The interaction concerned is n-σ^* as before, and independent of the main π-conjugation which is responsible for the planarity of the system.) This explanation is supported by some calculations on mono- and dihydroxyethylene; which also favour the Z-conformation (*142*) about the C–O bonds. The calculations are consistent with n-σ^*_{C-C} overlap being the source of the stabilisation of (*142*) over its E-isomer and since the σ^* antibonding orbital of a C=O bond is expected to be lower in energy, and thus to interact more effectively than that of C=C, this stabilisation is expected to be greater still in an acid or ester.

This preference of carboxylic acids and esters for the Z conformation about the =C–OR single bond is only the most familiar of a number of related conformational preferences, though few have been studied in any detail. A similar preference is observed for imidate esters, which generally exist in conformation (*143*) [236].

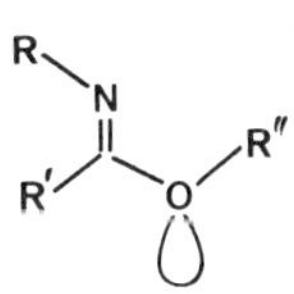

143

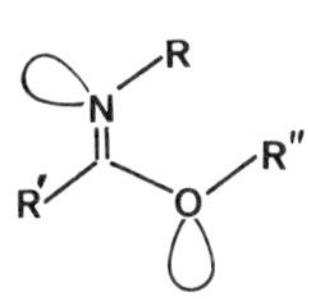

144

In some cases the form (*144*) with the Z-configuration across the C=N double bond also, is preferred, in spite of the strong 1,3-interaction between R and R″ in this structure, suggesting that there is a significant driving force for the electronegative substituent on carbon to be *antiperiplanar* to the nitrogen lone pair. This preference is not uncommon [237–239], though it is not general, for systems with potential n_N-σ^*_{C-O} interactions [240]. But there is evidence that it is stronger for more electronegative substituents on carbon. Thus equilibrium in the system (*145*) is close to 1 : 1 [239]; while the hydroximoyl chlorides (*146*), where the interaction concerned is n_N-σ^*_{C-Cl}, exist at equilibrium almost exclusively in the Z-form [241].

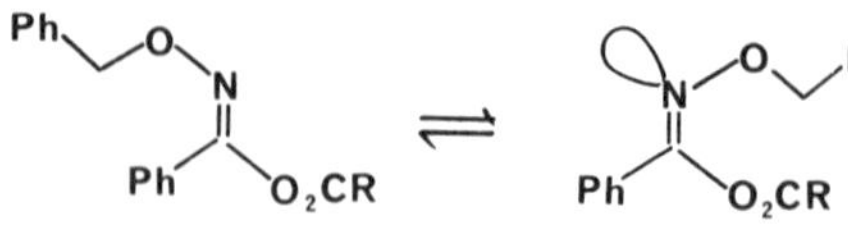

E 145 Z

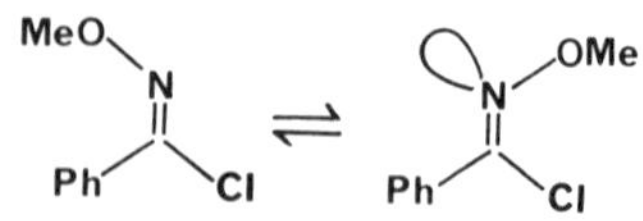

E 146 Z

The n-$\sigma^*_{C=N}$ stabilisation would be expected to be largest in the $C{=}N^+$ system, but though calculations on protonated acetamide [241a] predict the structure (*147*) there is little experimental evidence. The crystal structure of protonated dimethylacetamide indicates that the conformer (*148*) with the oxygen lone

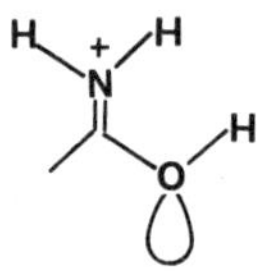

147

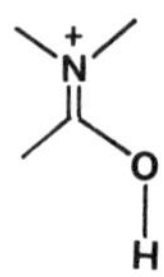

148

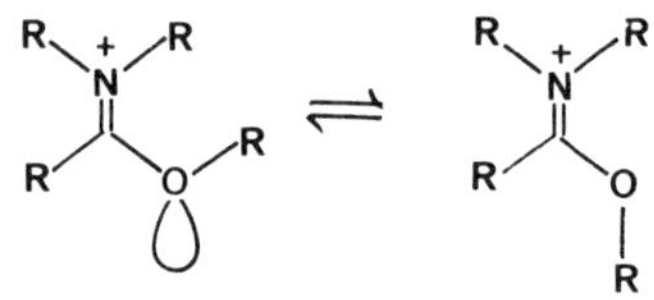

149

pair *antiperiplanar* to the $C{-}CH_3$ bond is preferred in the solid; but this is not a fair test because the hydrogen atom concerned is involved in H-bonding to the chloride counter-ion, so crystal packing forces may be dominant. There is also no clear evidence about the preferred conformations of the corresponding esters (*149*) [242].

There are, however, scattered pieces of evidence for distinct conformational preferences of OMe groups bound to sp²-hybridised carbon in olefinic and aromatic systems which are difficult to interpret except in terms of stereoelectronic effects. Examples are compounds (*150*) and (*151*), shown by

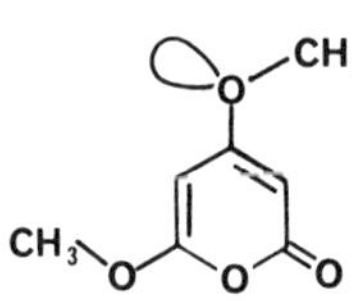

150

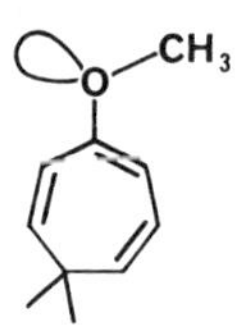

151

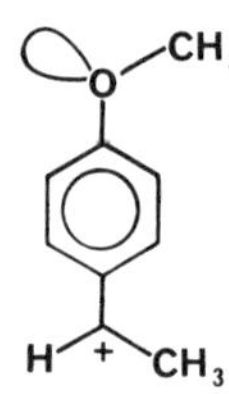

152

nuclear Overhauser effect measurements to adopt the conformations shown [243]; and the even more remarkable case of the benzylic carbonium ion (*152*), shown by NMR to exist predominantly (61 %) in the form with the methyl groups *syn* to each other [244]. All these cases, with the exception of the 6-MeO group of (*150*), appear to prefer the conformation with the lone pair on oxygen *antiperiplanar* to the bond to the more electronegative atom, however remote.

In all these cases of preferred conformations about $RO{-}C{=}$ bonds there is an inference, though no strong evidence, that the origin is a stabilisation by n-σ^*_{C-X} overlap of the form with a non-bonding pair of electrons *antiperiplanar*

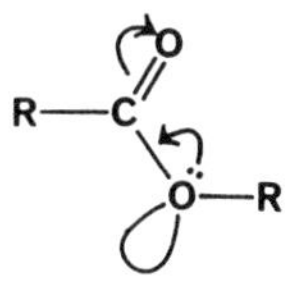

153

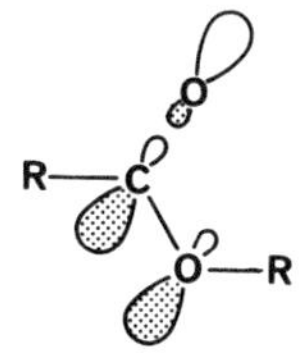

154

to a C=X bond. Because the non-bonding electrons with π-symmetry are already involved in π-bonding (*153*), only the low-energy σ-lone pair is available, and has the correct symmetry, to overlap with, for example, the antibonding σ^*_{C-O} orbital of the C=O bond (*154*). It might be expected, therefore, that the effect would be weaker than the anomeric effect in acetals; but the preference of esters for Z (*140*) over E (*141*) conformations by as much as 3 kcal mol^{-1} [234, 245] suggests that this is not the case. Part of the explanation is the geometry of the RO−C system, which is more favourable for overlap, compared with n-σ^*_{C-O} in an acetal, for three reasons: the shorter =C−OR bond; the O=C−O−R dihedral angle, which is 0°, and thus fixed in exactly the right conformation for overlap; and the fact that the n and σ^*_{C-X} orbitals are parallel in the sp^2-hybridised system.

One final piece of evidence, which is nicely consistent with the n-σ^*_{C-X} overlap explanation, is the observation that the clear preference for the structure with the oxygen lone pair *antiperiplanar* to the C=O bond disappears in chloroformate esters (*155*) [246–248] and is probably reversed in fluoroformates (*156*) [249].

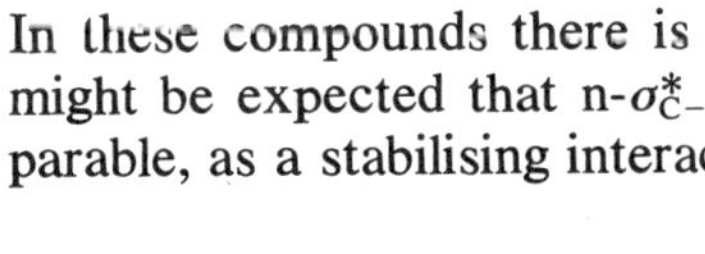

155

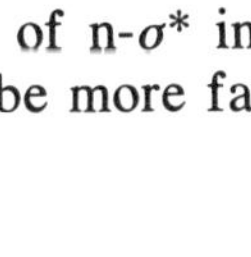

156

In these compounds there is a choice of n-σ^* interactions available, and it might be expected that n-σ^*_{C-F} would be more favourable, and n-σ^*_{C-Cl} comparable, as a stabilising interaction.

b. The Gauche Effect

Stabilising overlap between non-bonding electron-pairs and the antibonding σ^*-orbitals of adjacent bonds, as described above for systems showing the anomeric effect, also provides the most satisfying explanation of the gauche effect in systems R−X−Y−R′, with adjacent heteroatoms. Since the effect is expected to be strongest for the lowest-lying σ^* orbitals, we consider first the molecule FOOF. With only one ligand at each end of the O−O bond steric effects are simplified, and would be expected to favour the *antiperi-*

planar geometry. In the gas phase, however, this system shows a dihedral angle close to 90° [250], which is that expected if n-σ^*_{C-F} overlap is a significant stabilising factor in both directions (*157*). The very short O−O (1.22 Å) and very long O−F bonds (1.58 Å) are consistent with this double π-contribution to O−O bonding, and to the weakening of the O−F bonds caused by populating σ^*_{O-F}.

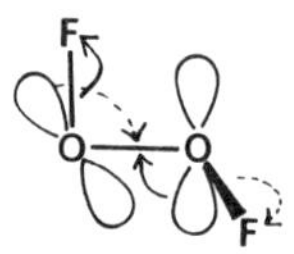

157

The same perpendicular geometry is observed not only for compounds like FOOF, with strongly electronegative terminal atoms, but also for H_2O_2 and other peroxides, disulphides and hydrazines (see above, page 32). All prefer conformations with the (π-type) non-bonding electron-pairs orthogonal to each other, or *antiperiplanar* to the bond on the adjacent atom. The generalisation which accounts most economically for these preferences is that compounds R−X−Y−R′, with two adjacent heteroatoms X and Y, tend to adopt conformations in which π-type lone pairs on X are coplanar with the Y−R′ bond (and similarly for n_Y and X−R). The stabilisation from n-σ^*-overlap which this allows will be greatest for bonds (X−R, Y−R′) to the most strongly electronegative substituents R, R′). For the conformation where the non-bonding electron-pairs are parallel no antibonding orbital is available to overlap with either, and only electron-pair repulsion is possible.

This generalisation can be extended to systems X−C−C−Y, which also show evidence for a gauche effect in certain cases (see above, page 30). Here there are no lone pairs on the central carbon atoms, only bonding electrons. But a σ-bonding electron pair is also a potential electron donor, given sufficient demand, and the geometry is always favourable for σ-σ^* overlap in staggered conformations (*158*).

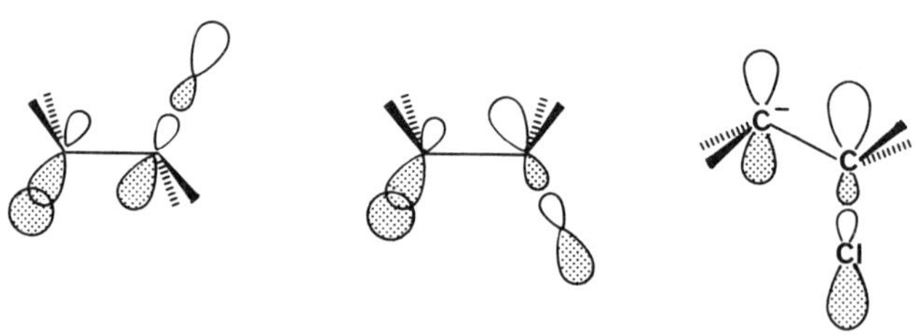

158 *159* *160*

(Epiotis et al. [251] have given a formal proof that *anti*-overlap (*158*) is more efficient than *syn*-overlap (*159*); which is also less likely in practice because it requires an unfavourable eclipsed conformation.)

Thus across any σ-bond each of the three different staggered conformations allows up to six different n-σ^* or σ-σ^* interactions. Of these we may expect σ-σ^* interactions involving only C−H and C−C bonds to be very weak (though there is a school of thought. [252, 253] which attributes the threefold barrier to rotation of ethane largely to such σ_{CH}-σ^*_{CH} interactions). The most effective donor will be the lone pair of a carbanion, n_{C^-}, followed by the sequence

$n_N > n_O > \sigma_{CC}, \sigma_{CH} > \sigma_{CX}$ (where X is O, N, S, Hal);

while the most effective acceptor, after the vacant π-type orbital of a carbonium ion [254], will be the lowest energy antibonding orbital $\sigma_{CHal}^* > \sigma_{CO}^* > \sigma_{CN}^* > \sigma_{CC}^*, \sigma_{CH}^*$ (for a more complete list see Ref. [225a]). The strongest stabilising interactions will be between the most effective donors and the most effective acceptors, and in systems where a powerful donor is adjacent to a relatively low energy antibonding orbital this single interaction will be dominant. An extreme case would be a carbanion with an adjacent C-heteroatom bond: an α-chlorocarbanion, for example, is estimated to prefer the eclipsed conformation (*160*) by a massive 15.7 kcal mol^{-1} [255, 256]; but the nonbonding electrons of O^- also have a significant effect on the strength of the adjacent C−H bond in CH_3−C^- [257–259].

Compared with n-σ^* effects of this sort, σ-σ^* interactions will always be relatively weak. The strongest will be expected to be between σ_{CC} or σ_{CH} and σ_{CF}^*: the two σ_{CH}-σ_{CF}^* interactions possible in the gauche conformation of FCH_2CH_2F, for example, can explain the observed preference for this form under certain conditions; and the fact that the tendency to adopt gauche conformations is smaller in other XCH_2CH_2F systems, and falls off as the sum of the electronegativities of X and Y decreases (see above, page 34).

This picture of a molecule as a composite of donor lone pairs and bonds, and acceptor bonds, was first clearly enunciated, and applied to stereochemical questions, by Epiotis et al. [251], and leads to a generalisation broad enough to include both the anomeric effect and the gauche effects observed in systems R−X−Y−R′ and X−C−C−Y.

"In the absence of overriding effects of other sorts, the preferred geometry about each bond of a molecule will be the one which maximises the interactions between the best donor lone pair or bond, and the best acceptor bond. The orientation allowing maximal n-σ^* and σ-σ^* interaction is antiperiplanar."

Or, more briefly:

There is a stereoelectronic preference for conformations in which the best donor lone pair or bond is antiperiplanar to the best acceptor bond.

C. Stereoelectronic Effects on Reactivity: The Kinetic Anomeric Effect

1. Introduction

So far we have discussed specifically stereoelectronic effects on structure: the anomeric effect as usually understood is a thermodynamic effect, concerned with the ground state properties of molecules. But the most exciting recent developments in this area involve stereoelectronic effects on reactivity. Because *transition states* may be more sensitive to stereoelectronic effects than the ground states of molecules, these effects may in principle be magnified when they operate on rates of reactions. This is a rapidly developing field, and conclusive evidence is often difficult to obtain, so that our conclusions will generally be preliminary ones. But when the available evidence is put together there is already an impressive body of results which are difficult to explain except in terms of a 'kinetic anomeric effect' [260].

a. The Use of sp^3-hybrid Lone Pairs

In our discussion of the stereoelectronic basis of the anomeric effect (Section B. 6, above) we used the formally correct description of the non-bonding electron pairs of oxygen, in terms of two canonical molecular orbitals, one with σ and one with π-symmetry. Of these the latter is considerably higher in energy, and can thus, to a first approximation, be considered as the one involved in bonding interactions with available vacant orbitals of higher energy.

Organic chemists, including those working on the kinetic anomeric effect, almost invariably prefer to use the alternative description in terms of two equivalent sp^3-hybridised lone pairs; for a variety of reasons, of which familiarity and ease and consistency of representation are important. We now have to decide, in the light of our conclusions about the status of lone pair electrons, and the stereoelectronic basis of the anomeric effect, whether this practice is "justified, unjustified but harmless, or actually misleading" (page 39). Only if it is positively misleading would there be sufficient reason to depart from accepted practice.

In terms of canonical molecular orbitals the n-σ^* interaction, which is the stereoelectronic basis of the anomeric effect in a system $RO-CH_2-X$, is strongest in the orthogonal conformation (*161*), in which the dihedral angle between the π-type lone pair and the C−X bond is 0°.
In practice the conformations of systems stabilised by the anomeric effect are

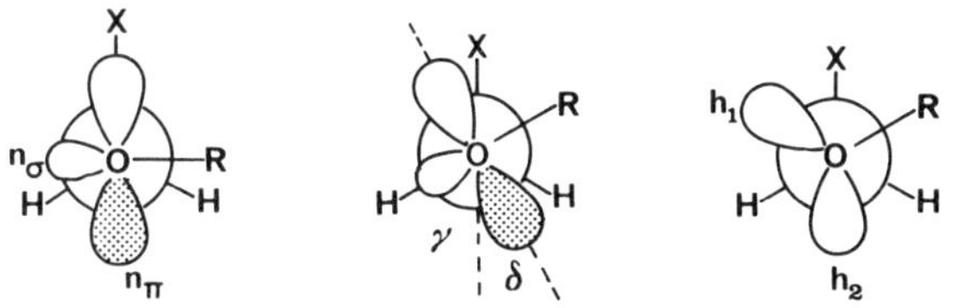

161 162 163

determined by a balance of competing effects, and the result is generally close to the gauche conformation (*162*). Now whereas in (*161*) n_π-σ^*_{C-X} overlap is maximised, (and n_σ-σ^*_{C-X} overlap is zero), the picture for (*162*) is more complicated. The strongest overlap is still n_π-σ^*_{C-X}, because n_π is closer in energy to σ^*_{C-X}, and because its orientation is closer to optimal (angle $\delta = 30°$) for overlap with σ^*_{C-X} than is that of n_σ (angle $\gamma = 60°$). Nevertheless n_σ-σ^* overlap is formally possible in the gauche conformation, and the toal n-σ^* stabilisation will be the sum of contributions from the two sources.

Compare this picture with the alternative description in terms of hybrid lone pairs (*163*). Here n-σ^*_{C-X} overlap is taken to be with the *antiperiplanar* sp^3-hybrid lone pair (h_2), which is made up of a small contribution from the s-orbital (σ-symmetry) and three times as much p-contribution (π-symmetry). This is clearly closely similar to the description in terms of canonical lone pairs, where n-σ_{C-X} overlap is mainly n_π-σ^*_{C-X}, with a small contribution from n_σ-σ^*_{C-X}. In fact a semi-quantative analysis of the differences between the canonical and localised molecular orbital descriptions of n-σ^* stabilisation (Appendix, page 135) shows that for all qualitative purposes the two can be regarded as indistinguishable.

We are therefore entirely justified in using sp^3-hybrid lone pairs in discussing the anomeric and kinetic anomeric effects, and will do so for the remainder of this book. (The generalisation arrived at the end of Section B does not depend on a particular description of lone pairs.)

b. The Stereoelectronic Basis of the Kinetic Anomeric Effect

The familiar frontier orbital description of the S_N2 reaction between a simple alkyl halide and a nucleophile Nu involves an interaction between a non-bonding electron pair in the highest occupied molecular orbital (HOMO) of the nucleophile and the lowest unoccupied MO (LUMO) of the halide (*164*), which is the antibonding σ^*-orbital of the C−X bond.

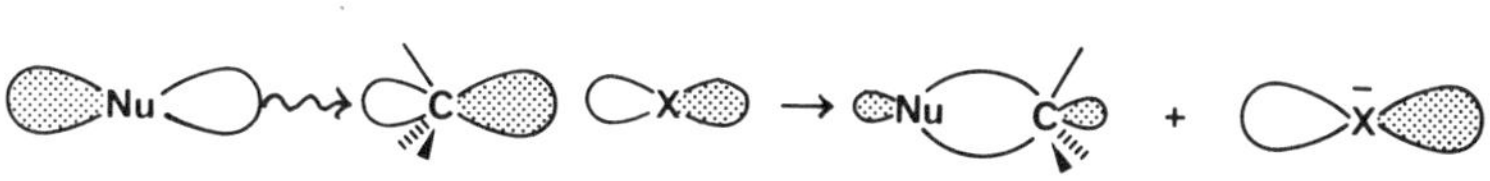

164

Overlap is bonding for the approach with σ-symmetry as in (*164*). Thus a collision with this orientation can lead to population of σ^*_{C-X} by electron density from the nucleophile, and hence if the total energy of the system is sufficient, to the cleavage of the C−X bond and the departure of the leaving group (X^-) with its bonding pair of electrons. The Walden inversion is a logical consequence of this series of events.

There are obvious imilarities between the n-σ^*_{C-X} interaction depicted in (*164*) and the n-σ^*_{C-X} interaction which is the basis of the anomeric effect (*165*); and here too it might be expected that population of σ^*_{C-X} by electron den-

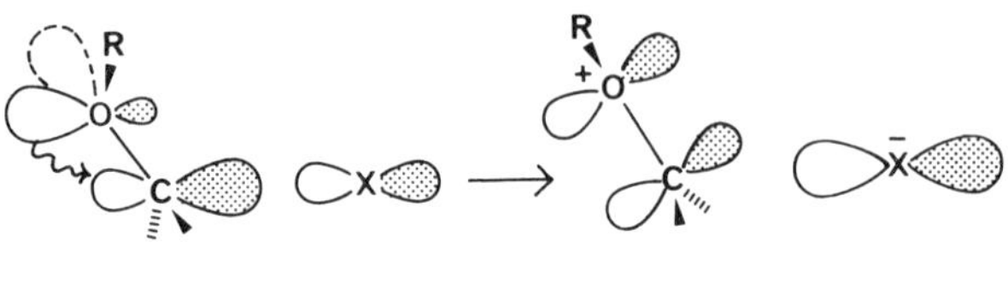

165 166

density from the non-bonding orbitals of the *adjacent* oxygen atom (also a HOMO-LUMO interaction, but in this case intramolecular, and with π-symmetry) could, if the total energy of the system is sufficient, lead to the cleavage of the C−X bond (*166*).

Even in the ground state the C−X bond is weakened by this n-σ^* interaction. This is clear from evidence of many sorts, including the lengthening of the C−X bond (see above, page 70); the shift in the ^{35}Cl nuclear quadrupole resonance in α-chloroethers [261]; the increase in the infra-red stretching frequency of the bond (apparent even for ν_{CH} (X=H) in extreme cases) [262]; the reduced barriers to rotation about the C-N bonds of axial N-glycosides [57]: and the increased shielding of the X nucleus observed in ther nmr spectra of many compounds [265a]. In all cases where there are conformational restrictions these effects are observed specifically for compounds with the gauche (*sc*) conformation about the C−O bond of the RO−C−X system. Furthermore it is clear, for example from the pattern of bond length changes, that the stronger the n-σ^*_{C-X} overlap (which is of course net stabilising for the gauche conformation of the ground state), the more the C−X bond is weakened.

What happens in the extreme case, where the donor-acceptor interaction is the strongest possible, is described by some calculations on the E1cb reaction of $FCH_2CH_2^-$, which is isoelectronic with FCH_2OH.

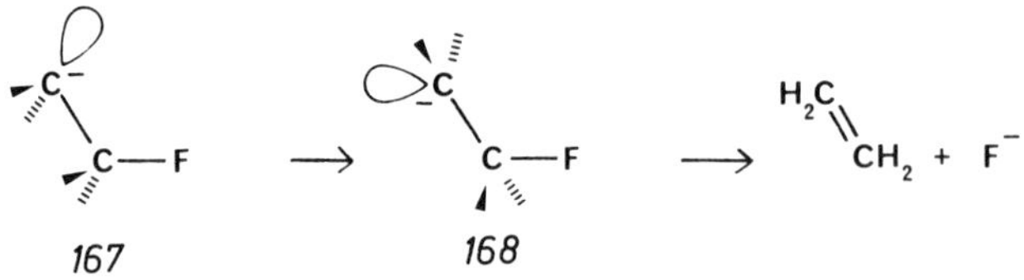

167 168

Both *syn* (*167*) and *anti* (*168*) isomers are unstable with respect to products, and full energy minimisation (4–31G basis set) leads directly to elimination: though the conformer with the lone pair and the C−F bond held *orthogonal* to each other is 'stable' [263]. By extending the C−F distance in small steps, but holding it fixed for individual calculations, a reasonable reaction path could be mapped out for the *anti* isomer (*168*), with the CH_2 group flattening out smoothly as the C−F bond breaks. Under the same conditions the calculated reaction path for the less stable *syn* isomer (*167*) starts with an abrupt, exothermic inversion of configuration, and is subsequently identical with that of the *anti*-isomer. Neither reaction involves any activation energy, and the trajectory of the departing fluoride ion is similar, in reverse, to that (*169*) calculated for the addition of hydride ion to formaldehyde by other authors [264].

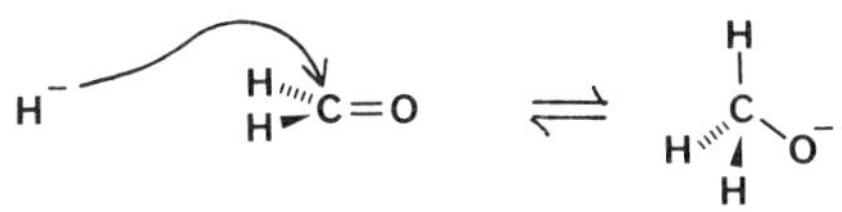

169

Thus both frontier orbital considerations and *ab initio* calculations support the idea that C−X bond cleavage in RO−C−X and related systems will be easiest in conformations in which lone pair orbitals on Y are coplanar with, and preferably *antiplanar* to, the C−X bond.

What do we expect to happen if this is not the case? In practice the most stable conformation of a system of this sort does have lone pair electrons *antiperiplanar* to the best leaving group, as a result of the anomeric effect. And even if it does not, because of some minor conformational restriction, reaction can still proceed *via* the favoured *antiperiplanar* conformation if this is reasonably accessible. But if the system is rigid, it is readily demonstrated that the kinetic anomeric effect may in principle be far larger than the anomeric effect on ground states.

Consider the elimination of X^- from CH_3OCH_2X. In the stable gauche conformation (*170*) elimination can proceed smoothly to generate the expected oxocarbonium ion (*171*) in its ground state configuration.

170 *171*

Elimination from the *antiperiplanar* conformation (*172*), however, will – if the conformation about the C−O bond is fixed – produce the perpendicular, high energy form (*173*) of the oxocarbonium ion in which π-bonding is prevented.

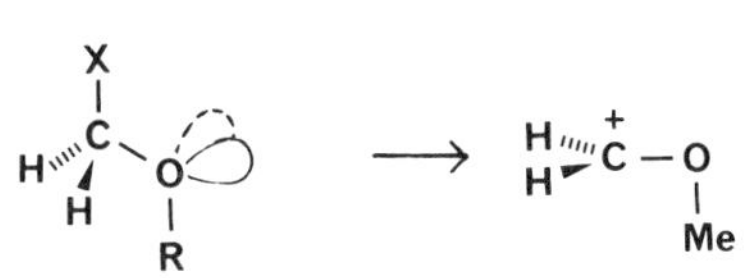

172 *173*

The difference in energy, δ, between the two ground state conformations (*170* and *172*) is small (typically 2–3 kcal mol^{-1}, as discussed in Section A. 2, above). But the difference in energy between the oxocarbonium ion products (*171* and *173*) and thus, to a good approximation, that (Δ) between the nearby transition states leading to them also, is very much greater (Fig. 19).

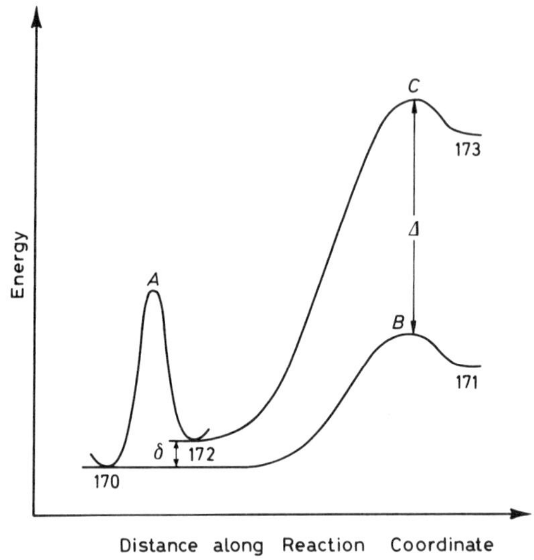

Fig. 19. Energy profile diagrams for cleavage of the C–X bonds of the (*sc*) (*170*) and (*ap*) (*172*) conformations of $ROCH_2X$, and for rotation about the O–C bond (curve A). Δ is the difference ($E_C - E_B$) in energy between the transition states (B,C) leading to the planar (*171*) and perpendicular (*173*) conformations of the oxocarbonium ion, and δ is the difference in ground state energies

This difference is roughly equal to the barrier to rotation about the $CH_2{=}\overset{+}{O}Me$ double bond, which has been estimated to be about 20 kcal mol^{-1} [265]. (Somewhat higher barriers for protonated formaldehyde, $CH_2{=}\overset{+}{O}H$, of 25.2 [266a] and 31.4 kcal mol^{-1} [266b] have been calculated by different authors.) An *extra* barrier of this magnitude is equivalent to an adverse rate factor of over 10^{14}, more than enough to prevent entirely any reaction which is observed under normal conditions for the gauche conformer. (The corresponding barriers for rotation about the $C{=}\overset{+}{O}H$ bonds of protonated acetaldehyde (27.9 kcal mol^{-1}) and formic acid (31.4 kcal mol^{-1}) [266b] show that very large effects are in principle possible for other acetals, and for orthoesters and their derivatives also.

In practice molecules are not completely rigid, and a thermodynamically reactive system held in a conformation corresponding to (*172*) may have enough flexibility to allow significant overlap between a lone pair on oxygen and the developing carbonium centre by the time the transition state is reached. But the magnitude of the kinetic anomeric effect is potentially so large that we must expect to observe large effects in suitable systems.

So far we have been discussing an extreme case, where two conformations (*170* and *172*) react entirely independently. If these are two conformations of the same molecule this corresponds to a situation where the rotational barrier between them is higher than the higher of the two transition states for C−X cleavage (see Fig. 19). But the general requirement for a significant stereoelectronic effect on reactivity (i.e., a faster cleavage of *170* than *172*) is only that the rotational barrier for the conversion of form (*172*), without a lone pair *antiperiplanar* to the C−X bond to the 'reactive' form (*170*), should be higher than the activation energy for cleavage of (*170*). If the conformational barrier is lower than this both conformers can react through the same transition state (*B*), and their relative reactivities will be determined by their ground state energies. If it is higher (but still lower than the transition state (*C*) for cleavage without stereoelectronic assistance), both forms can again react through transition state (*B*), but the rate determining step for the reaction of (*172*) will now be the conformational change to (*170*). The overall reaction of (*172*) will thus be slower than that of (*170*), as a direct result of the higher stereoelectronic barrier to direct cleavage, though the observed rate bears no direct relationship to the size of this barrier.

In principle this state of affairs might be achieved either in a system where the conformational barrier is high, which in practice means it must be more or less rigid: or by making the cleavage step very fast. In this latter case stereoelectronic effects on the reactivity of conformationally mobile systems might become significant.

It is not unusual for acyl transfer reactions to involve mechanisms where the rate determining step involves a proton transfer in the thermodynamically favourable direction, between a general acid or base and the appropriate site of a tetrahedral addition intermediate. The slow step of a reaction of this sort is generally not the proton transfer itself, but the diffusion of the catalyst to meet the high energy intermediate. In cases of this sort the diffusion step is by definition slower than the breakdown of the intermediate, which involves the cleavage of carbon-heteroatom bonds, to starting materials or products. Now the apparent activation energy for such a diffusion process in water is of the order of 3 kcal mol^{-1}, similar to barriers to rotation about the single bonds of simple alkanes. So in tetrahedral intermediates involved in the reactions of carboxylic acid derivatives the barriers to the cleavage of C−X bonds could be lower than the barriers to rotation about single bonds.

In such a situation the direction of cleavage will depend on the low energy pathways available to the intermediate *in the conformation in which it is first formed.* Such pathways must be stereoelectronically favourable, and the transition states of the cleavage pathways will be by definition close in energy to the intermediate, and depend much less on the stability of the products. Thus product ratios can be subject to *stereoelectronic control* [235], which is a special case of kinetic control. And since the conformational barriers which separate the channels to different products can amount to several kilocalories, the effects on product ratios should be readily observed. (A stereoelectronic, or other, barrier of 1.3 kcal mol^{-1} is enough to lead to a product ratio as large as 9 : 1 at 25°.)

Stereoelectronic effects on the *reactivity* of conformational isomers of such mobile systems, on the other hand, are expected to be considerably smaller, and much more difficult to observe. The requirement for one conformer to be less reactive than another for stereoelectronic reasons, as discussed above, is that the rotational barrier E_A between conformational isomers should be greater than the activation energy for the cleavage of the 'reactive' form (*170* in Fig. 19). This means $E_A > E_B + \delta$ (Fig. 19). Since the stabilisation of the 'reactive' form by the anomeric effect (δ) will be of the order of 1–2 kcal mol^{-1}, and the barrier to rotation may be no more than about 3 kcal mol^{-1} in an acyclic system, any stereoelectronic effect on reactivity is unlikely to be more than 1–2 kcal mol^{-1}. And since systems reactive enough to break down faster than they can reach conformational equilibrium can only be generated as reactive intermediates from more stable precursors, their reactivity cannot normally be measured directly. Rate constants derived indirectly are less reliable, and depend on assumptions about the mechanism. So looking for stereoelectronic effects on the *reactivity* of conformationally mobile systems is not a promising approach.

Thus estimates of the magnitude of stereoelectronic effects, which require measurements of reactivity, require systems which are conformationally more or less rigid. But it ought to be possible to observe the operation of stereoelectronic effects on product distributions in mobile systems also, if these can be generated in conformationally different forms which break down faster than they are interconverted. The rest of the book describes the search for stereoelectronic effects using these two approaches.

2. Reactions of the Acetal Group

a. Acetal Hydrolysis

In principle the simplest way to demonstrate stereoelectronic control [235] of acetal cleavage would be to show that the axial and equatorial isomers of a given tetrahydropyranyl acetal react at different rates under the same conditions. An axial acetal (*174*) can cleave smoothly, maintaining maximal n-σ^* overlap throughout the process, whereas if the equatorial isomer (*175*) is to cleave it must do so without assistance from the lone pairs on oxygen, because antiperiplanar to the exocyclic bond in this case is only a σ-bonding orbital of the ring.

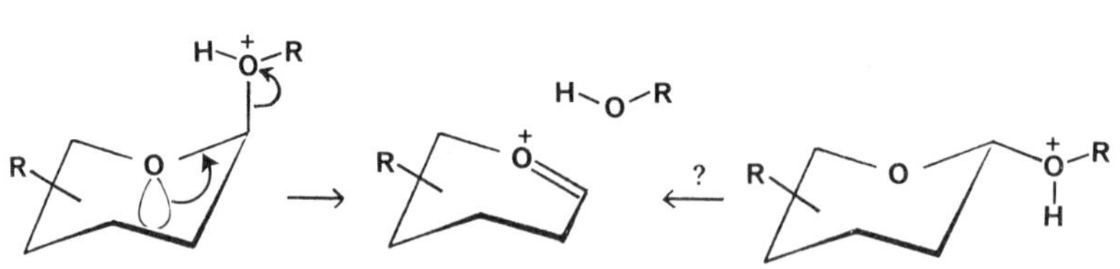

174 176 175

It is clear (for example from Fig. 19) that there is no chance of observing differential reactivity between anomeric acetals such as (*174*) and (*175*) if they are interconvertible, because ring flipping allows (*175*) to cleave with stereoelectronic control via (*174*). (Note that in this case the equatorial isomer is likely to react *more* rapidly, because the anomers cleave to give the same oxocarbonium ion (*176*), and thus via transition states of very similar energy. See discussion on page 87).

So a conformationally fixed system is required. Glucose derivatives might appear to be suitable, because of the four or five equatorial substituents, but they too are evidently not rigid enough because methyl α-glycosides, with axial leaving groups, are hydrolysed more slowly than the equatorial anomers [267]. In the case of aryl glycosides the α-isomers are hydrolysed a few times faster than the β-anomers [268], but the effect is small, and complicated by the initial proton transfer equilibrium:

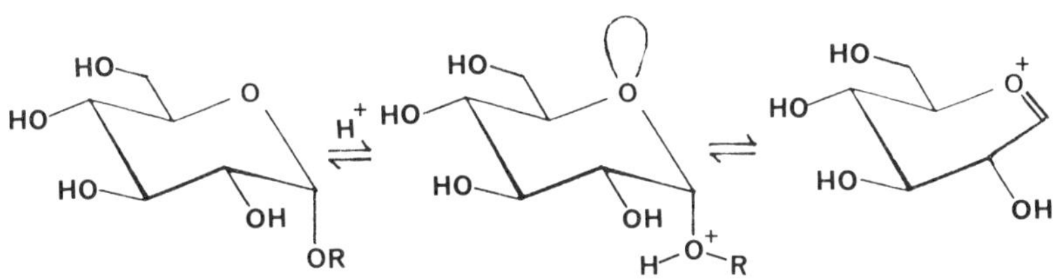

177 178

It is probable that the relative stabilities of α and β-glycosides are reversed in the conjugate acids (e.g. *177*) (the reverse anomeric effect; see page 15). This could mean that the reactivity of the conjugate acid of an axial compound (e.g. *177*), which is no longer stabilised by the anomeric effect, is increased: on the other hand the basicity of the axial OR group will be decreased by the same effect, so that less conjugate acid is present at a given pH. (And *vice versa* for the equatorial anomer.) Clearly the situation is complicated, but the two effects are in opposition, and could cancel out even if they are substantial. We have no way at present of knowing whether this is so.

Another difficulty with the acid-catalysed reactions of alkyl tetrahydropyranyl acetals is that cleavage of the ring C–O bond can provide a mechanism for isomerisation, which could clearly vitiate results with a rigid system (e.g. *179*⇌*180*).

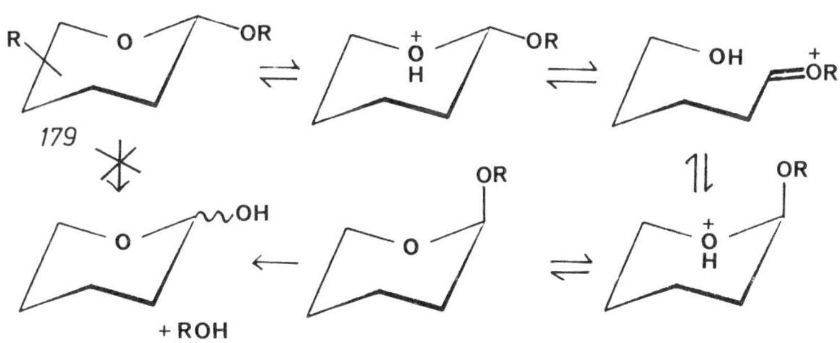

179

180

Although it has been shown that the primary mechanism of anomerisation of methyl glucosides involves exocyclic C–O cleavage [269], this applies to reactions going with stereoelectronic control. If in the case of a rigid equatorial system (*179*) the cleavage of the exocyclic C–O bond is indeed very slow, then it is to be expected that the alternative anomerisation mechanism, *via* (*180*), would prevail (if there is no stereoelectronic barrier to this pathway).

A solution to both these problems is to use aryl tetrahydropyranyl acetals (*181*) with good leaving groups ArO^-. These are electronically too unsymmetrical for the ring acetal C–O bond to cleave at a significant rate, even in acid;[7] but reactive enough in the other direction for spontaneous cleavage (*181*→*182*) of the exocyclic C–O bond to be observed [270, 271]. Thus the hydrolysis of acetals of this type may be studied in the absence of acid catalysis.

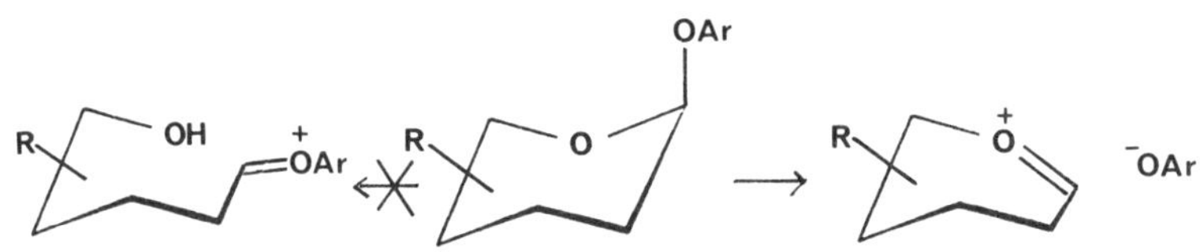

181 182

The first test of these ideas [272] used the oxadecalin acetals (*183* and *184*, Ar = p-nitrophenyl).

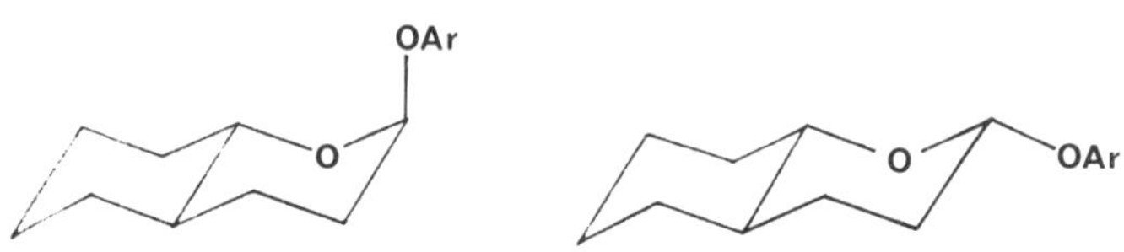

183 184

There is a substantial body of evidence based on related systems for stereoelectronic control in E2 reactions. The Hofmann elimination of trimethylamine from (axial) 3α and 6β-trimethylammoniocholestanes, for example,

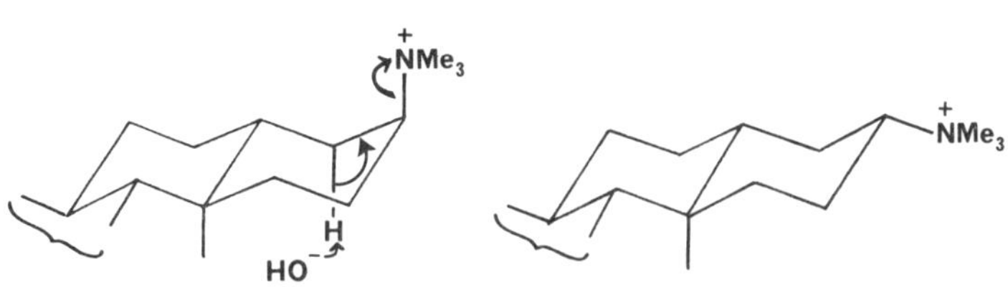

185 186

[7] Even nitrophenoxymethyl chlorides, $ArOCH_2Cl$, are unreactive compounds. As the oxygen atom becomes more electronegative α-haloethers eventually come to resemble methylene chloride in reactivity.

is much faster than that from the equatorial (3β and 6α) isomers (compare part structures (*185*) and (*186*) [273, 274]. So in this case the *trans* ring junction clearly holds the conformation of the A-ring of (*186*) rigid enough to prevent easy elimination *via* other conformations.

This is not the case for the oxadecalin (*184*, Ar = p-nitrophenyl). This acetal is hydrolysed three times faster than the axial anomer (*183*) [272], and though there are minor differences in kinetic parameters for the reactions of (*183*) and (*184*) it is clear that the two compounds are cleaved *via* closely similar transition states, and that their relative reactivity is determined almost exclusively by their ground state energies.[8]

The likely explanation is that the equatorial isomer (*184*) reacts by way of a conformation close to the boat (*187*), in which a lone pair on the ring oxygen is *antiperiplanar* to the exocyclic C–O bond:

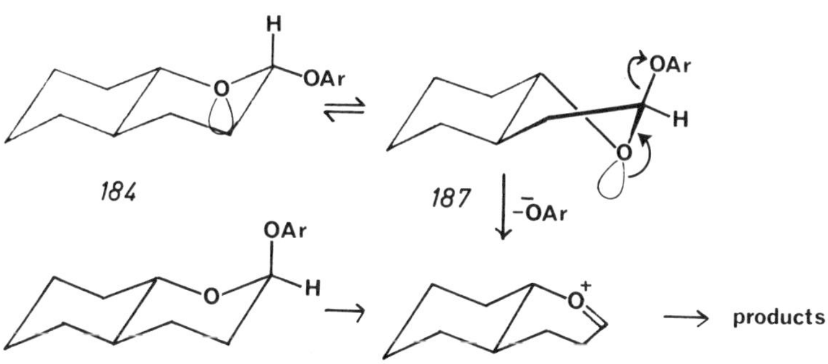

183

This is much more favourable than would be the corresponding conformational change of (*186*), which could put a β-hydrogen atom *antiperiplanar* to the leaving group and so in principle allow the E2 reaction to proceed. In the trimethylammonium compound the Me_3N^+ group itself (A-value ~ 5 kcal mol^{-1}, like that of t-Bu [275]) is large enough to lock the ring in the chair conformation; whereas in the acetal the anomeric effect favours the boat over the chair conformation, so that the free energy change for the equilibrium (*184*) ⇌ (*187*) will be quite small; perhaps 2–3 kcal mol^{-1}, compared with a free energy of activation for C–O cleavage of some 25 kcal mol^{-1} [270]. (The

[8] A similar result has been found for the acid hydrolysis of the anomeric acetals (i) and (ii) [275a].

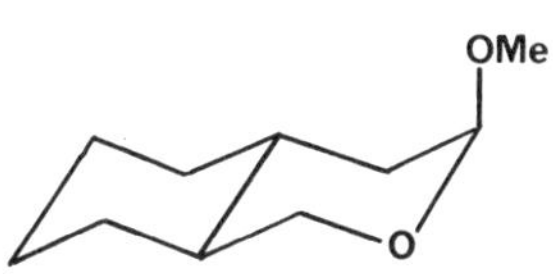

(i)

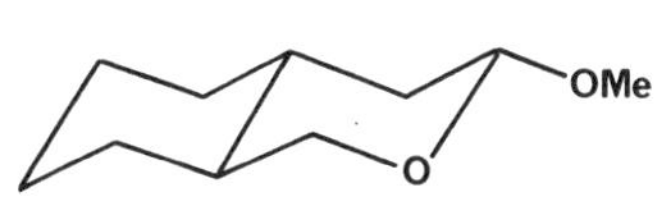

(ii)

twist-boat is in fact the most stable conformation of several related compounds (see pages 12 and 30)). Thus is appears that the oxadecalin system is not rigid enough for stereoelectronic effects to control reactivity.

The first successful demonstration involved the axial-equatorial pair (*188*) and (*189*), where the conformation of the equatorial isomer (*189*) is fixed by the ring junction [276]. The axial anomer (*188*) is hydrolysed almost 400 times

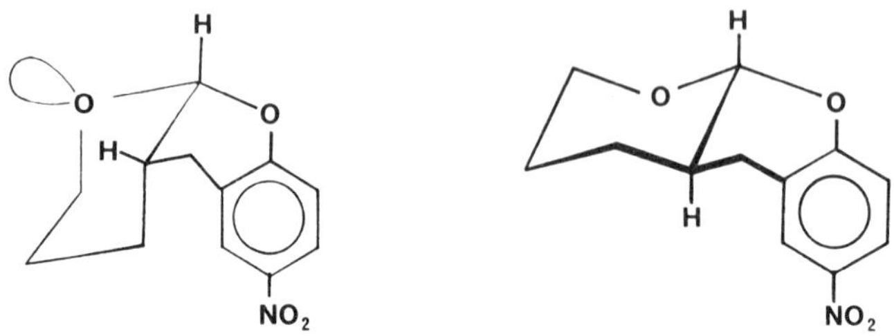

188 *189*

faster than the equatorial compound (*189*) in acid [276, 277], but spontaneous hydrolysis is very slow for both compounds. Nevertheless the axial isomer is hydrolysed slightly faster, although it is the more stable isomer. It is probable that a much greater difference in reactivity is masked by a change in rate determining step in the hydrolysis of the axial compound (*188*). The initial cleavage product (*190*) can revert rapidly to starting materials by intramolecular addition of ArO^- to the oxocarbonium ion, and has to be attacked by external water before it can go on to products.

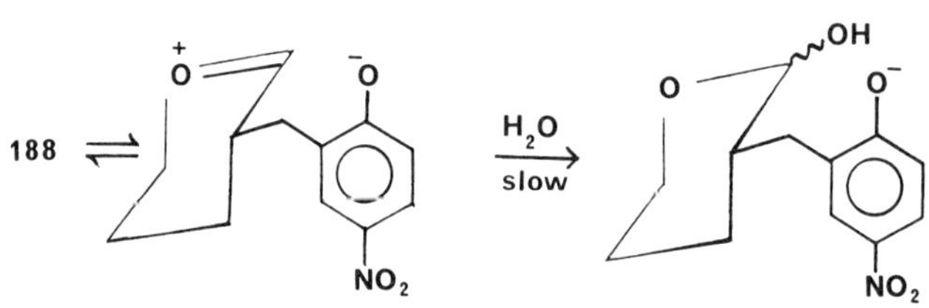

190 *191*

This second step appears to be rate determining: k_{H_2O}/k_{D_2O}, for example, is 1.8 at 100° for (*188*), although there is no significant solvent deuterium isotope effect on the hydrolysis of the equatorial anomer (*189*), for which C–O cleavage is presumably rate determining as usual. As a result the axial compound is some 10^4 times less reactive than expected (e.g. from the rate of hydrolysis of *183*). If the argument is correct the stereoelectronic factor, which alone is responsible for the very low reactivity of the equatorial isomer (*189*), can be estimated to be about 7 kcal mol^{-1}; substantial, though much smaller than the maximum possible value of some 20 kcal mol^{-1} suggested above (page 82).

The tricyclic system (*188* vs *189*) is not ideal because the main rate difference is concealed; and also because reaction does not go to completion but to an equilibrium mixture of (*188*), (*189*) and (*191*). A clearer demonstration of stereoelectronic control is found in the related system (*192*) [278].

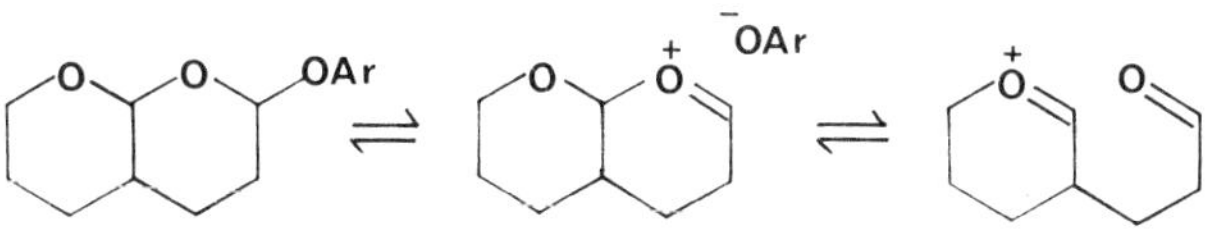

192 *193* *194*

Here the departure of the aryloxide leaving group would generate an acetal (*193*) with such a good (aldehyde) leaving group that it is unlikely to enjoy a significant lifetime before breaking down to (*194*) and products. The cleavage of the two C−O bonds is thus expected to be concerted, so that the departure of the leaving group is affected by the electron-donating capability of the remote oxygen atom. As before, this depends on the geometry of the ring junction: in isomer (*195*), with a *cis* ring-junction, this oxygen can act as a π-donor to assist the cleavage of the second C−O bond; whereas in isomer (*196*), with a *trans* ring-junction, it cannot, and will presumably act by inductive electron-withdrawal to inhibit the formation of (*193*).

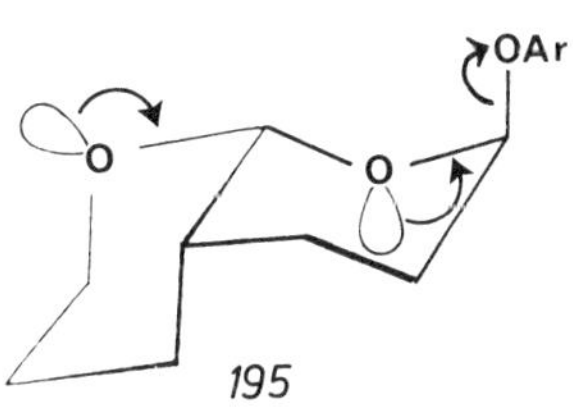

195

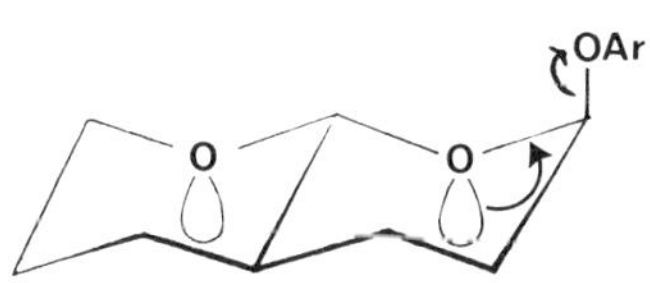

196

In fact (*195*, Ar = p-nitrophenyl) is hydrolysed 1380 times faster than the compound (*196*) with the *trans* ring junction. The enthalpy of activation is less favourable for the cleavage of (*196*) by 7.2 kcal mol^{-1}, suggesting again that a stereoelectronic barrier of about 7 kcal mol^{-1} is imposed by the particular geometrical restraints associated with the *trans* ring junction in systems of this type.

The largest effect observed to date involves the still more rigid system (*197*) [279]. This compound is hydrolysed over 10^{10} times more slowly than

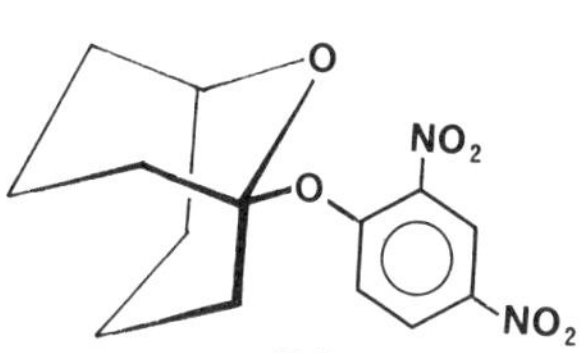

197

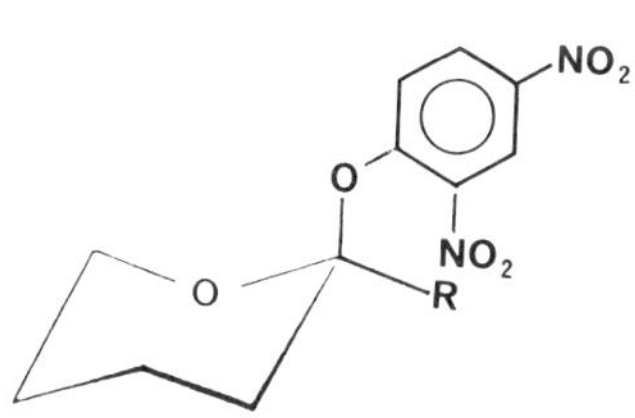

198

the corresponding axial 2,4-dinitrophenoxy acetal (*198*, R = Me; which is far too reactive to prepare[9]). The stereoelectronic barrier in this case is therefore of the order of 14 kcal mol^{-1}, suggesting that our estimate of 20 kcal mol^{-1} for the limiting value of the stereoelectronic barrier to acetal cleavage, in a notional completely rigid system, is not unreasonable.

A particularly graphic demonstration of the stereoelectronic effect in the early stages of acetal cleavage is the pattern of bond length changes in the ground states of axial and equatorial compounds like (*197*) and (*198*). Figure 20(a) shows how the length of the exocyclic C–O bond (x) increases with increasing electron-withdrawal in the Ar group for a series of axial 2-aryloxytetrahydropyrans (*199*). This bond lengthening is accompanied by a less regular shortening of the endocyclic C–O bond (n); evidently the same effect noted above (page 53) for the corresponding bond of α-chloroethers and α-pyranosyl chlorides. It has been suggested that plots like these map out the early stages of the reaction pathway for cleavage of the acetal group, which of course involves more extreme changes in the same direction (page 80).

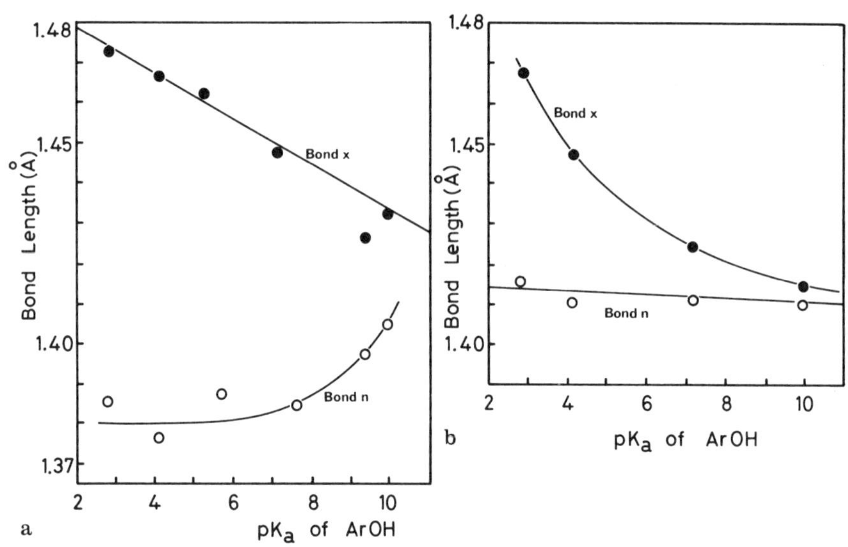

Fig. 20a and b. The contrasting patterns of bond length changes at the acetal centres of **a** axial *(199)* and **b** equatorial 2-aryloxytetrahydropyrans, as a function of the basicity of the aryloxy group. The exocyclic C–O bonds (x, closed circles) are always longer than the endocyclic bonds (n, open symbols)

[9] The comparison is therefore based on an extrapolation. The half-life of (*198*. R = H) can be estimated to be about 1 second, on the basis of the good linear free energy relationship established for the hydrolysis of a series of less reactive 2-aryloxytetrahydropyrans [271]. This is 3×10^9 times shorter than that of (*197*) under the same conditions. And (*198*, R = Me) will be considerably more reactive than (*198*, R = H).

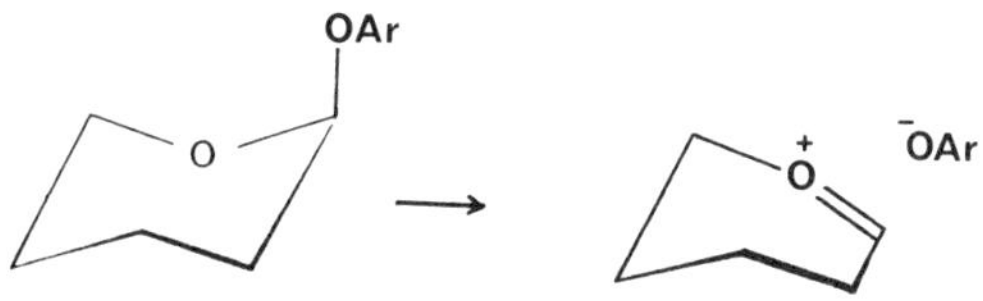

199

The pattern of bond length changes observed for the corresponding equatorial compounds is quite different (Fig. 20b). For the phenyl acetal the two bond lengths are quite similar, though the exocyclic bond is significantly shorter than that of the corresponding axial compound. The introduction of electron withdrawing substituents into the aromatic ring leads to a lengthening of the exocyclic C−O bond (*x*) as before. But the pattern is clearly different, and in particular the associated shortening of the endocyclic C−O bond (*n*) has disappeared. This evidence too is consistent with a transmission of electronic effects through the acetal centre in the axial compounds which is closely linked to the pathway for cleavage of the exocyclic bond, and is interrupted in the equatorial system.

The hydrolysis and formation of five-membered ring acetals (1,3-dioxolanes) present an intriguing problem in stereoelectronics. Consider the usual mechanism (Scheme) for the reaction of the simplest formaldehyde acetal.

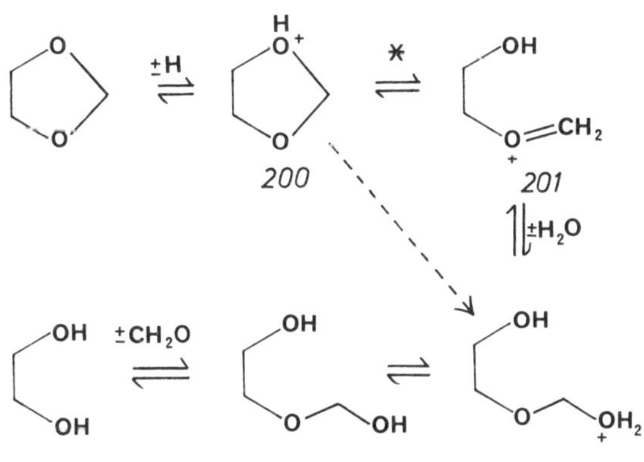

Scheme

The problem arises in the step (*) in which the C−O bond of the ring is made and broken. The bond in question is *anticlinal* to both lone pairs on the other ring oxygen (*200a*) and cannot get *antiperiplanar* to either.

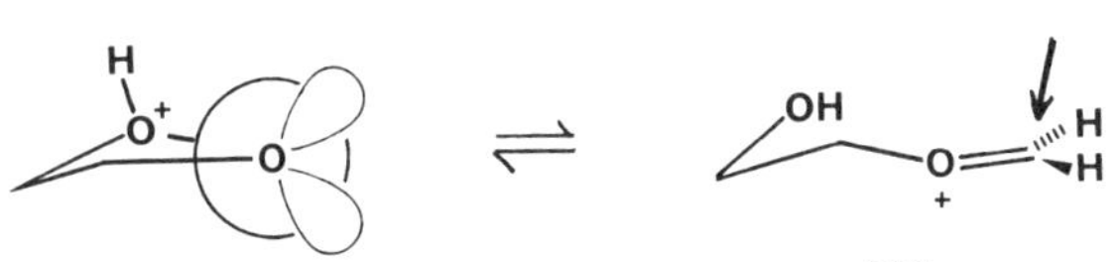

200a 201a

Similarly the recyclisation of the oxocarbonium (*201*) is stereoelectronically difficult (it is a 5-*endo-trig* reaction [280]), because the OH group cannot reach the preferred trajectory (*201a*) for nucleophilic addition to the C=OR group.

What is certain is that the reaction goes, and that reactivity does not appear to be significantly reduced for stereoelectronic reasons. Where there is a choice between five- and six-membered ring formation, as in reactions of aldehydes with glycerol and sugars, the five-membered ring is often the kinetic product, even in cases where the six-membered cyclic acetal is thermodynamically more stable [281]. And though the hydrolysis of five-membered ring acetals is often much slower than that of comparable acyclic compounds, this can generally be attributed to the reversibility of the ring opening step.[10] So either the mechanism shown in the Scheme is wrong, and reaction goes by an alternative pathway which gets round the apparent stereoelectronic barrier. Or the mechanism is correct, and the barrier, if any, is overcome.

The most likely alternative mechanism involves nucleophilic attack by water on the protonated acetal (*200b*, corresponding to the dashed arrow in the Scheme) [282]. This would lead directly to ring-opened products and does avoid the oxocarbonium ion intermediate. It is also in line with recent work on formaldehyde acetals with good leaving groups [283, 284], which are displaced by a range of nucleophiles, including water; and has led to the suggestion that oxocarbonium ions $R\overset{+}{O}=CH_2$ derived from formaldehyde are in fact too unstable to exist for a significant length of time, thus forcing the reactions to be concerted [284, 285]. Furthermore, the kinetic parameters which indicate that a molecule of water is involved in the transition state for the hydrolysis of some 1,3-dioxolanes (negative entropies of activation, and normal solvent deuterium isotope effects) are as consistent with the concerted mechanism (*200b*) as with rate determining addition to (*201*) [286].

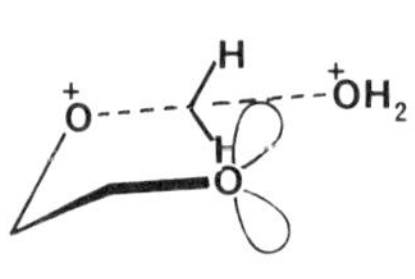

200b

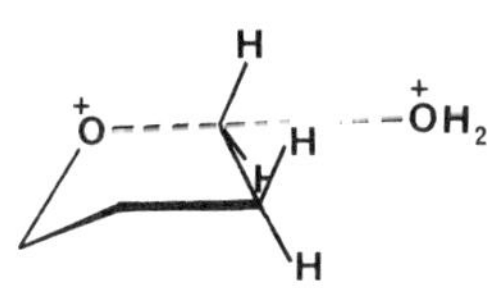

200c

The case for the concerted mechanism (*200b*) is therefore strong; but it is not entirely convincing. The hydrolyses of acetals derived from ethylene glycol – the least likely to have hydration of the oxocarbonium ion rate determining – are very clearly A1 reactions, with positive or near-zero entropies

[10] The intramolecular addition of OH in (*201*) can readily be faster than the external addition of water, which thus becomes the rate determining step: as shown in relevant cases by the slower hydrolyses of aldehyde and ketone acetals, compared with formaldehyde derivatives; reactions which would be faster if C–O cleavage were rate determining.

of activation [287] similar to those observed for the hydrolysis of similar acyclic acetals (Table 21). And the reactions are much slower for formaldehyde acetals compared with acetaldehyde or acetone derivatives. Furthermore, the spontaneous hydrolysis of benzylidene catechol (*200d*) [288] almost certainly involves rate determining addition of water to the oxocarbonium centre:[11]

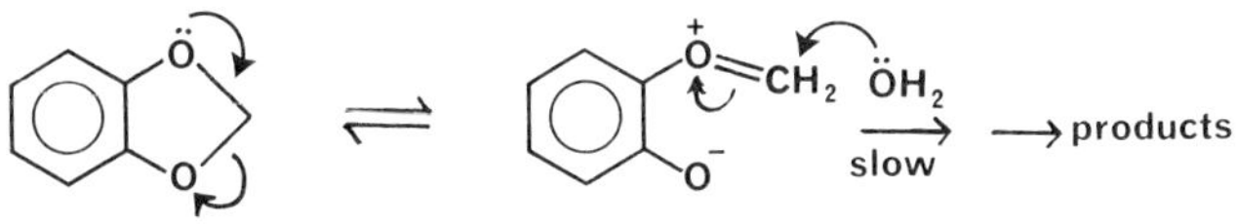

200d

and the more stable Schiff-base intermediate (*200e*) can actually be observed in the hydrolysis of some oxazolidines [289].

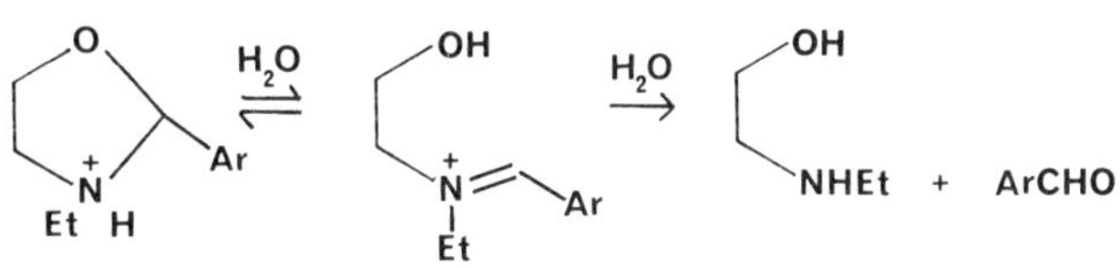

200e

In fact the concerted mechanism is not a complete solution to the basic stereoelectronic problem either. The original assumption was that the *anticlinal* lone pairs on the ring oxygen (*200a*) do not overlap sufficiently well with the σ^*-orbital of the C−O bond to be broken to assist cleavage efficiently. But a *concerted* displacement at an acetal centre is also dependent on electron-donation from the 'spectator' oxygen [283, 284]: S_N2 reactions at acetal centres are much faster than displacements of similar leaving groups from CH_3, and the α-deuterium isotope effect is consistent with a high degree of S_N1-character for S_N2 reactions of $MeOCH_2$ systems [283, 284]. Similarly the hydrolysis of a cyclic acetal, slow though it may be, is still much faster than the cleavage of a cyclic ether. This is difficult to explain if both involve concerted displacements (compare *200b* and *200c*).

The conclusion seems inescapable that there is significant assistance from the two *anticlinal* lone pairs of (*200a*) for the cleavage of the other C−O bond. This is consistent with some calculations by Gorenstein, which show that two such *anticlinal* lone pairs are slightly more effective than one *antiperiplanar* lone pair in assisting P−O cleavage in a pentaoxyphosphorane [338]. And

---

[11] For the same reasons discussed above (footnote, page 92) for reaction through (*201*): except that it is even clearer in this case that the addition of external water will be slower than that of neighbouring aryloxide. The authors [288] suggest general acid catalysis of ring opening by water; but the proton transfer involved never becomes thermodynamically favourable, so would not help.

applies whether the cleavage of the C−O bond is concerted or unimolecular. Most likely it is unimolecular for derivatives of ketones and aldehydes. For the hydrolysis of 1,3-dioxolane itself the entropy of activation (Table 21) is a few units smaller, which might be an indication that a molecule of solvent is more specifically involved, in the sense indicated in (*200b*), for formaldehyde derivatives. If so, this involvement is no greater than in the reactions of comparable acyclic acetals, and it is not a substitute for assistance by the lone pairs on the donor oxygen.

Table 21. Kinetic Parameters for the Hydrolysis of 1,3-Dioxolanes Derived From Formaldehyde, Acetaldehyde and Acetone [287]

	k_{H+} (1.mol^{-1}s^{-1})	$\Delta H^{\neq}$ (kcal mol^{-1})	$\Delta S^{\neq}$ (e.u.)
	2.65×10^{-6}	24.9	−0.6
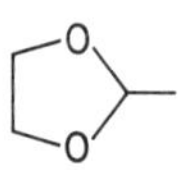	1.36×10^{-2}	21.6	+5.6
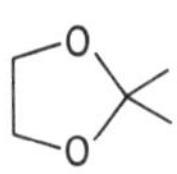	1.44×10^{-1}	20.9	+7.9

b. Acetal Formation

Insofar as acetals are generally made by way of the same oxocarbonium ion that is the first intermediate in their hydrolysis, their formation is expected to be subject to the same stereoelectronic effects. Large effects are not predicted for reactions involving a given (planar) oxocarbonium ion (e.g. *202*), because the transition states for addition of a nucleophile will be early, and close in

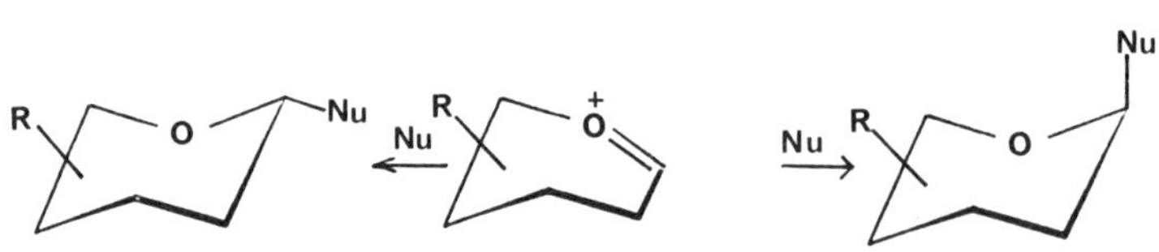

202a 202 202b

energy to the intermediate. The maximum difference in energy between the alternative pathways (*202*→*202a* or *b*) will therefore presumably not be attained until the new bond is fully formed, and will correspond, in the example above, to the anomeric effect favouring (*202b*) over (*202a*). Thus product ratios are expected to favour isomers with lone pairs *antiperiplanar* to the new C−Nu bond.

The available experimental evidence is consistent with this expectation. The oxadecalin acetals (*183*) and (*184*) (page 86), which are formed by heating a mixture of the corresponding acetates with the phenol in benzene, give the thermodynamic mixture of products, with the axial isomer predominating by about 3 : 1 [277]. So too does the cyclisation of the hydroxyacetal (*203*) [290].

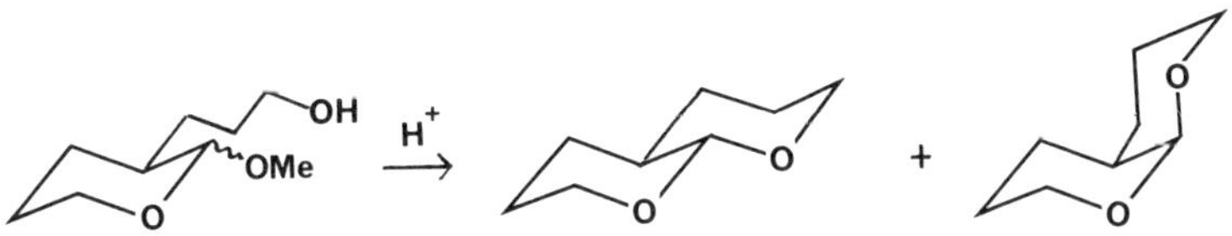

203

On the other hand the more rigid compound (*204*) is cyclised exclusively to the *cis*-acetal (*205*):

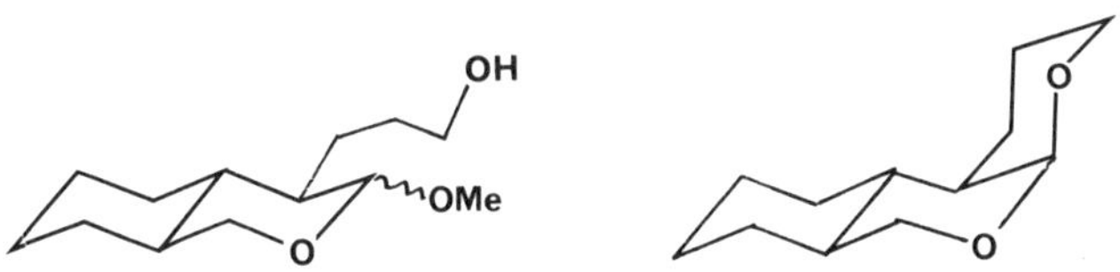

204 *205*

Heating the mixture of acetates (*206*) with *p*-nitrophenol in benzene, also gives the axial acetal (*207*) exclusively [277].

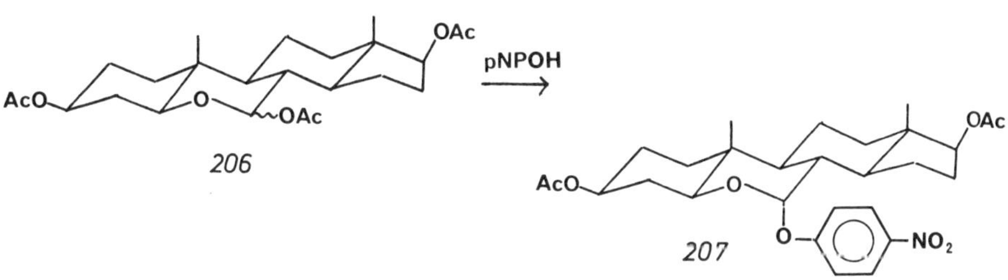

206 *207*

Similarly the *cis* acetal (*188*) is the initial product of the reaction between dihydropyran and the *o*-hydroxybenzyl bromide (*208*), which probably involves

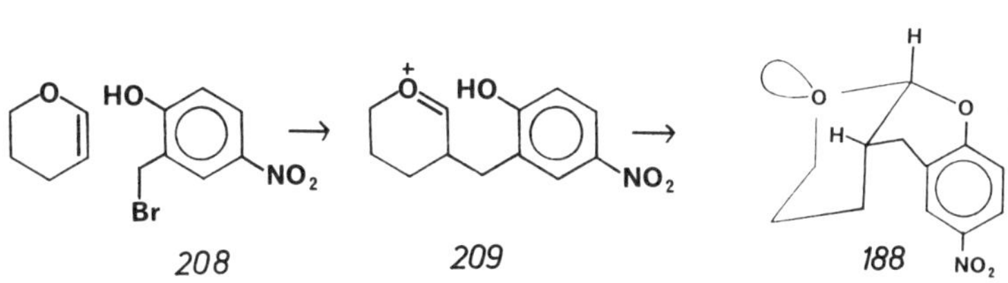

208 *209* *188*

the oxocarbonium ion (*209*) as an intermediate [276] (though a cycloaddition mechanism would also account for the observed stereochemistry). And the related intramolecular transfer of hydride ion to the C=OR centre of (*210*) [291] gives the product (*211*) with the side chain equatorial.

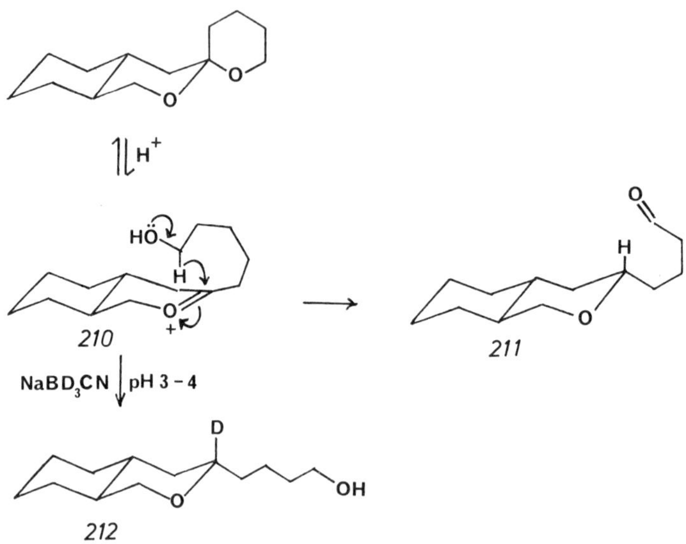

This is not a consequence of the intramolecularity of the hydride transfer reaction because reduction by cyanoborohydride also gives the equatorial isomer (*212*) of the corresponding alcohol [291].

The results in all these examples are consistent with addition of the nucleophile to generate a lone pair *antiperiplanar* to the bond being formed: which results in stereospecificity if the system is conformationally sufficiently rigid.

c. Acetal Oxidation

We saw above (page 77) that an *antiperiplanar* lone pair which is a sufficiently powerful donor (in, for example, an alkoxide anion) weakens even a C−H bond. Calculations [256–259] suggest that a lone pair on neutral oxygen has a similar, though much smaller effect; and this would undoubtedly be enhanced in the transition state for a reaction like that of the primary alcohol group of (*210*), in which electrophilic attack on the C−H bond leads to hydride transfer.

Apart from the Cannizzaro [292], and perhaps the Eschweiler-Clark reaction [293], hydride transfer from an acetal (*213*) requires a relatively strong electrophile (X^+), and the evidence suggests that reaction normally occurs only when lone pairs on *both* acetal oxygens are *antiperiplanar* to the bond being broken.

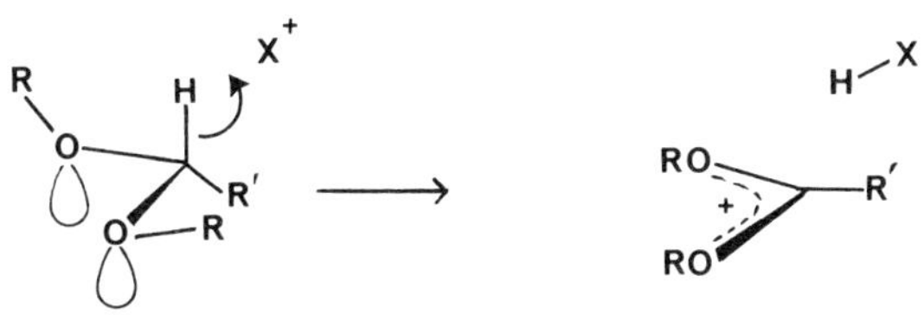

213

This is to be expected, in view of the evidence (page 89) that an oxygen atom which cannot act as an *n*-donor for conformational reasons will act exclusively by inductive electron-withdrawal to destabilize the developing oxocarbonium ion, and thus reduce the rate of its formation. Acetals are generally oxidised more rapidly than comparable ethers, implying that both oxygens do assist C−H cleavage, as in (*213*).

The original evidence came from work on the oxidation of sugars. Aldopyranoses with an axial C−H are oxidised by bromine more rapidly than the anomers [294]; and the CrO_3 oxidation of methyl pyranosides illustrates the point very clearly [295, 296]. Tetra-O-acetyl methyl β-D-glycopyranoside, which has two lone pairs *antiperiplanar* to the axial anomeric hydrogen atom in the ground state (*214*), is cleanly oxidised to the open-chain ketoester (*215*), no doubt via initial electrophilic attack on the C−H bond.

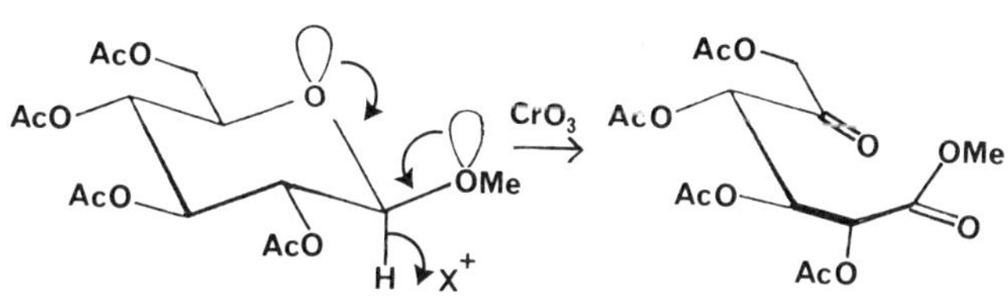

214 215

The axial anomer (*216*) on the other hand, cannot readily attain a conformation with a lone pair of the ring oxygen *antiperiplanar* to the equatorial C−H bond, and the only reaction observed is the slow oxidation of the methoxy-group [295, 296], as expected if the ring oxygen inhibits electrophilic attack on the acetal C−H bond.

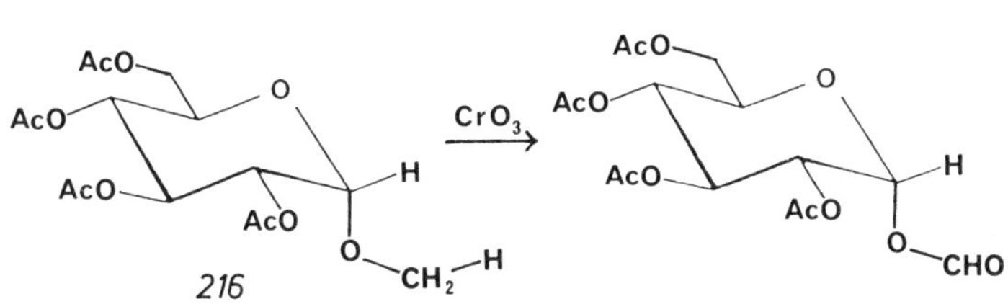

216

The same stereoelectronic factors are expected to operate generally for reactions involving electrophilic attack on the C−H bond of acetals, and other examples involve oxidation by N-bromsuccinimide [297] and ozone (see below), and the insertion reaction with dichlorocarbene [298]. *Nucleophilic* attack on the hydrogen atom, involving its removal as H^+, is also subject to stereoelectronic control, but now the n-σ_{C-H} 4-electron interaction, always destabilising in the ground state, will become more destabilising as reaction proceeds; hence the clear preference for metalation at the equatorial position in both acetals [299] and dithioacetals [300]. For example, the fixed-chair acetal (*217*), with an axial anomeric H, is not deprotonated at all under conditions where the anomer (*218*) is converted completely to the carbanion by butyl lithium [299].

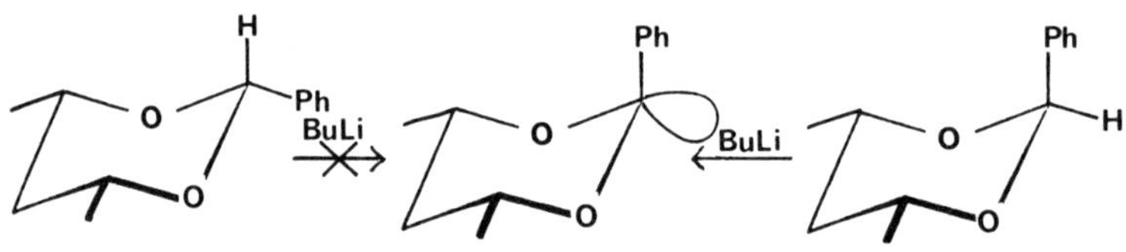

217 *218*

The most detailed studies of stereoelectronic effects on the oxidation of acetals involve the reaction with ozone [301–304]. The effects are very clear in this system, and the work was seminal in the development of the theory of stereoelectronic control by Deslongchamps and his group.

Initial attack by ozone on the C−H bond of an acetal is clearly electrophilic [305], and leads to the hydrotrioxide (*219*); which can be prepared as an unstable intermediate by carrying out the reaction in an inert solvent at low temperatures [306].

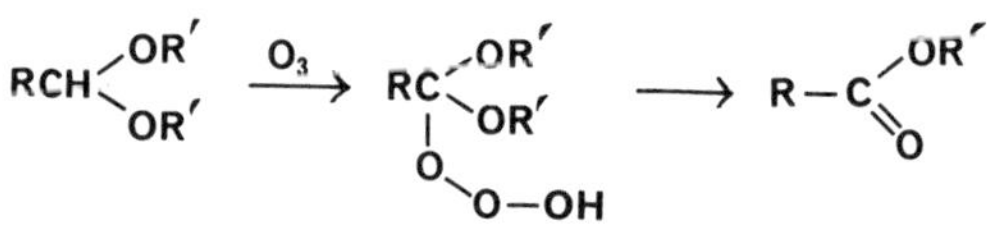

219

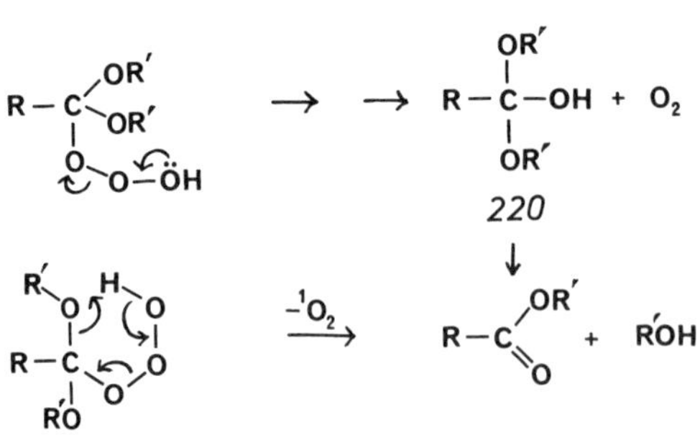

221

On warming the intermediate (*219*) decomposes, largely (but not exclusively) by an ionic mechanism [306], to give an ester. (The other products are the alcohol (R′OH) and singlet oxygen.) Possible pathways for the breakdown of (*219*) are by way of the hemi-orthoester (*220*) [303], or by a concerted mechanism (*221*) [306].
This part of the reaction is discussed in the following Section. For the moment we will consider specifically stereoelectronic effects on the initial stage, which involves attack by the electrophile on the C−H bond.

Assuming sp^3-hybridised lone pairs, the favoured gauche conformation of an acetal can exist in nine different conformations, (*222a–f* and the enantiomers of *222a, b* and *d*) [303].

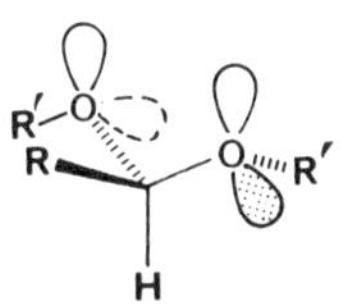

222a

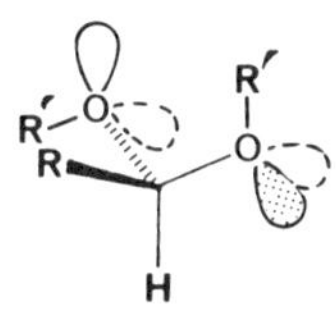

222b

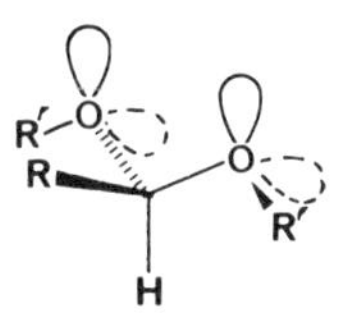

222c

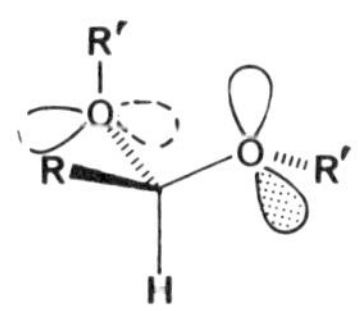

222d

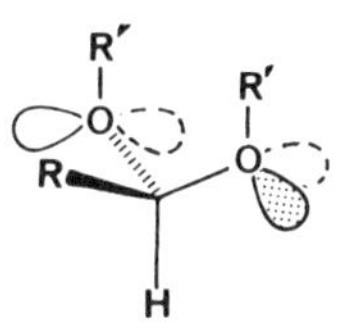

222e

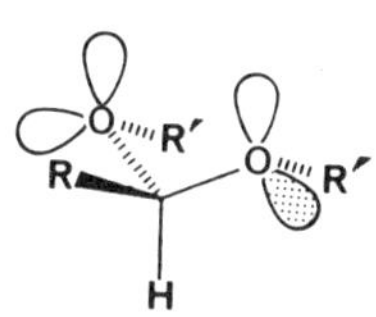

222f

Of these, conformers (*222a, c* and *f*) have lone pairs *antiperiplanar* to the C−H bond on both oxygens. Conformers (*222b* and *d*) have one, and conformer (*222e*) no such lone pair; so that the first three are predicted to be reactive, and the second three (*222b, d* and *e*) to be unreactive, towards electrophilic attack.

Deslongchamps and his co-workers have examined the reactivity towards ozonolysis of acetals with conformations corresponding to all six cases (*222a–f*), and find that these predictions are fulfilled exactly [303]. Thus methyl α-glycosides (*223*) which have an equatorial anomeric hydrogen atom, are unreactive, as they are towards other electrophilic reagents. The three conformations about the C−OMe bond correspond to conformers (*222d* and *e*), and to the

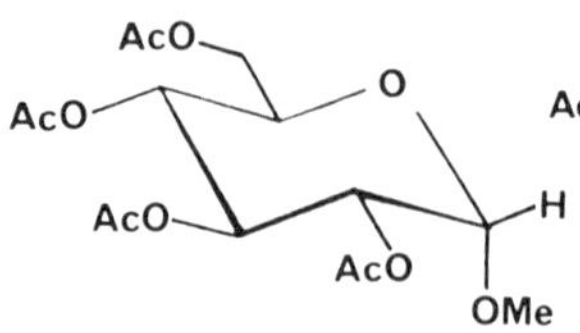

223

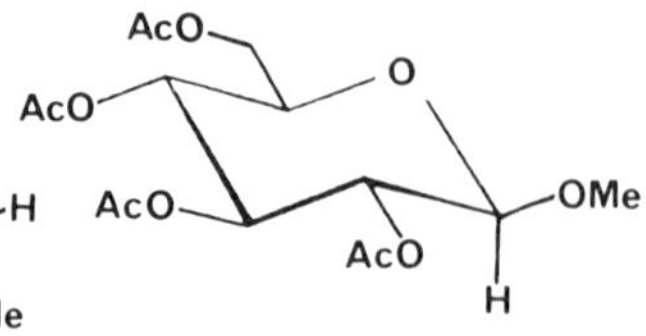

224

enantiomer of (*222b*): if any one of these were reactive then oxidation would be expected to occur. Since it does not, all three are unreactive, as predicted.[12]

Conformations (*222a, b* and *c*) correspond to the three possible conformations about the C−OMe bond of a *β*-glycoside, e.g. (*224*). We know already that conformation (*222b*) is unreactive, so the observed reactions of methyl *β*-glucosides like (*224*) must involve conformations corresponding to (*222a* or *c*). Of these the first is the preferred conformation of a methyl-*β*-glucoside, and probably the one principally involved in reaction, though a rigid model for this conformer has not been tested. A suitable rigid model for conformation (*222c*) is *trans*-1,8-dioxadecalin (*225*), which is readily oxidised

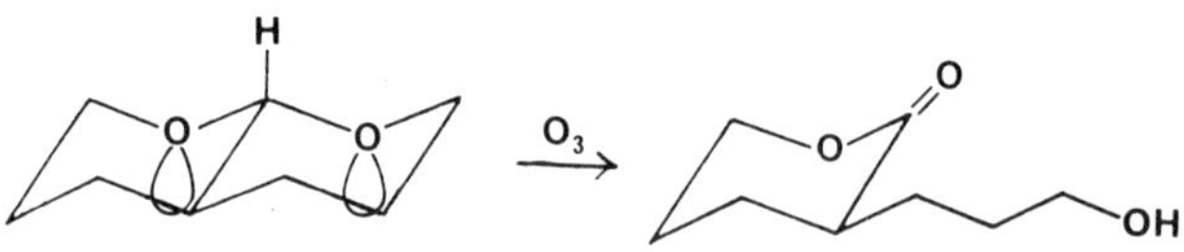

225

by ozone [303]; whereas the corresponding *cis*-decalin (*226*⇌*227*) which is a model for conformation (*222b*) is not.

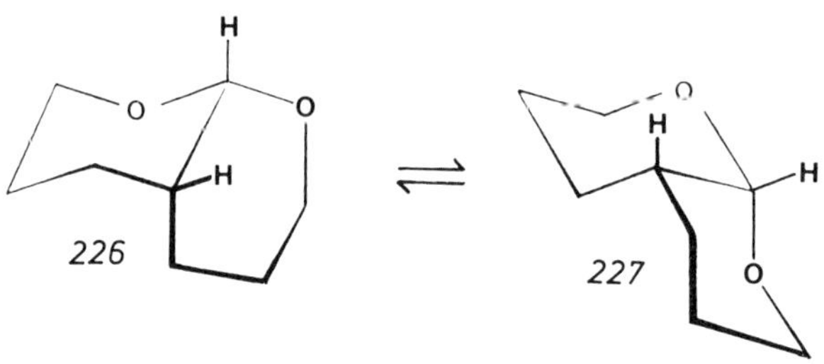

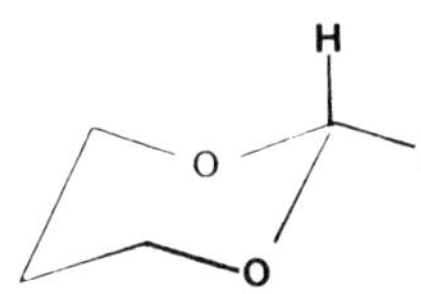

228

Finally, 1,3-dioxanes (*228*) which have the geometry corresponding to (*222f*), are readily oxidised by ozone, as predicted. (More readily, for example, than

[12] Note that the ground state ($^4C_1$) conformation of (*223*) is stabilised by the anomeric effect, so that flexible conformations of axial tetrahydropyran acetals are much less accessible than those of equatorial derivatives. This can explain why the hydrolyses of acetals like (*223*) and (*224*) go at similar rates, while their oxidation shows clear evidence of stereoelectronic effects.

acyclic acetals, which although not conformationally fixed, exist in the ground state in the unreactive conformation (*222d*) [303].

The breakdown of the hydrotrioxide intermediate formed by the attack of ozone on the C−H bond of the acetal group also appears to be subject to stereoelectronic control. This is apparent from the ozonolysis of tetrahydropyran acetals (*229*), which are converted specifically to acyclic hydroxyesters (*230*), rather than lactones [303].

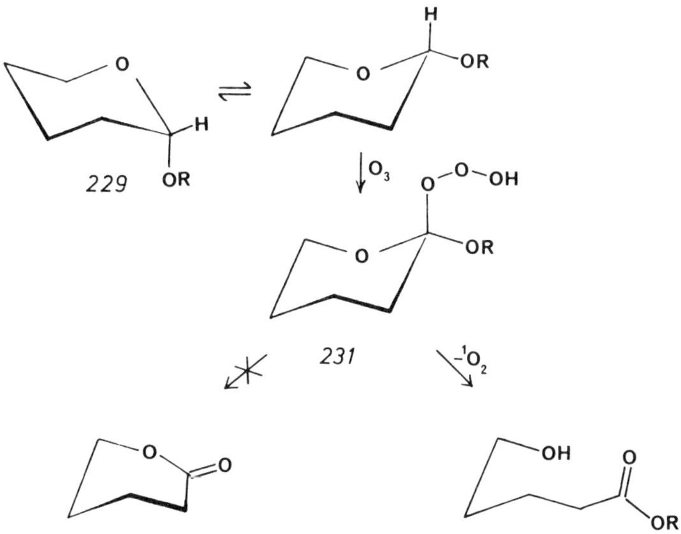

229 231 230

In the axial hydrotrioxide (*231*) (or derived hemiorthoester (*232*)), which will be that one produced by attack on the axial C−H, the ring oxygen cannot assist cleavage of the equatorial C−O bond; so that only the ring C−O bond can be cleaved with stereoelectronic control. (Assuming that if (*231*) or (*232*) are full intermediates, then breakdown to products is faster than ring inversion.) This is equally possible for the breakdown of the hemiorthoester (*232*), or the concerted breakdown of the hydrotrioxide (*233*).

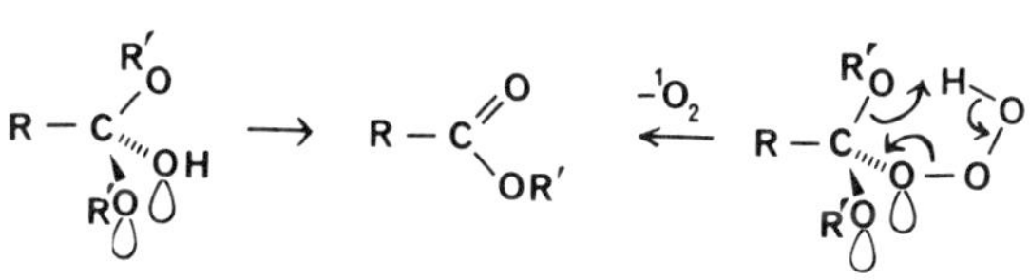

232 233

Considerations like these led to the investigations of stereoelectronic effects on the hydrolysis of orthoesters and their derivatives discussed in the following Section.

3. The Hydrolysis of Orthoesters

The hydrolysis of an orthoester (*234*) involves the initial cleavage of one of three C–O bonds, which may all be chemically non-equivalent. The stereoelectronic theory [234] predicts the selective cleavage of the bond or bonds

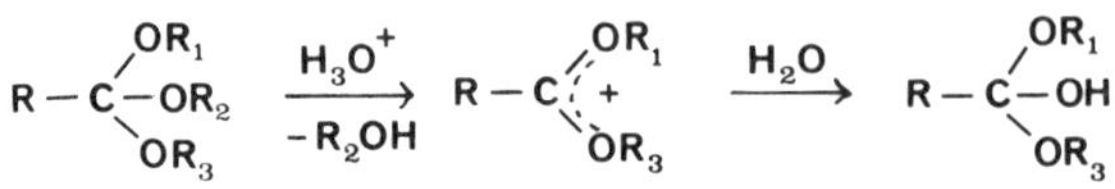

234 235

which can be *antiperiplanar* to lone pairs on both the remaining oxygen atoms, and the relevant experimental results are consistent with this expectation. Thus the hydrolyses of the three unsymmetrical esters (*236*, R = CH_3, CD_3 or Et), all lead to the same product (*237*), which has retained the OCH_3 group of the starting material [234, 307].

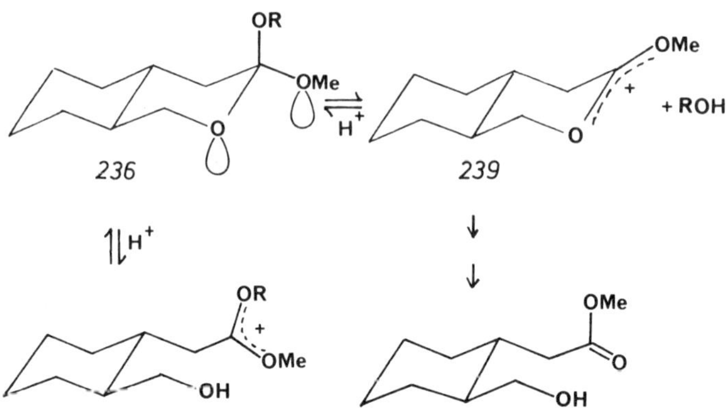

236 239 238 237

In principle the initial cleavage could have involved the ring C–O bond, but this would have led to the dioxocarbonium ion (*238*), which could not reasonably be expected (especially for R = CD_3) to lose ROH exclusively on hydrolysis. Loss of the OCH_3 group with stereoelectronic control could happen by way of a high energy flexible conformation (cf. the hydrolysis of the related equatorial acetal (*184*), page 86) but the extra barrier to reaction by this route is clearly too high. (Note that the orthoester (*236*) is midway, in terms of conformational rigidity, between the equatorial acetal (*184*) (page 86) which has a significant stabilisation from the anomeric effect in the non-chair conformations, and the equatorial trimethylammonium system (*186*), which is fixed firmly in the chair conformation and thus prevented from undergoing the Hofmann elimination (page 86)).

A full analysis of the conformational equilibria of a system like (*236*) has been presented by Deslongchamps [307]. Three staggered conformations are possible about each of the two exocyclic (OR and OCH_3) bonds, corresponding to conformations A1–A3 and E1–3 (see page 18, above), respectively; so that we have to consider nine possible distinct conformational isomers. Of these it is reasonable to ignore the three based on conformation A3 (page 18), with the R group over the ring, for steric reasons. Then of the set with conformation E2 about the equatorial C–O bond (*240*) only one (*241*)

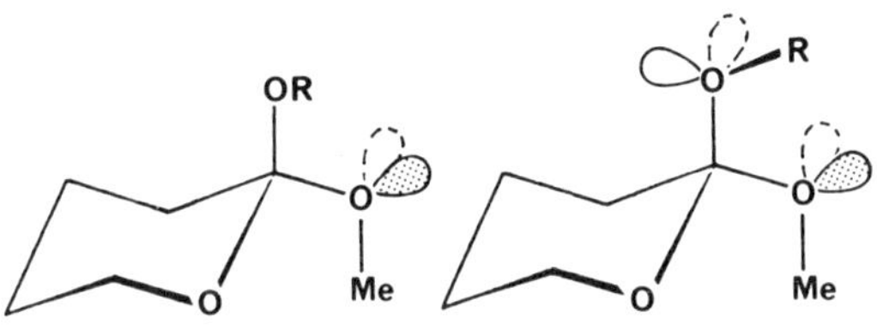

240 *241*

remains which has lone pairs on two (exocyclic) oxygens *antiperiplanar* to the third C–O bond.[13] This conformation is expected to react selectively by initial ring opening; which is not what is observed. It may be, as Deslongchamps et al. [307] suggest, that the entropy release expected for a reaction involving fragmentation to separate molecules favours the cleavage of an exocyclic C–O bond (though little entropy has in fact been released by the time the *transition state* for C–O cleavage has been reached). A more likely explanation involves the reversibility of ring opening. The neighbouring OH group of the dioxo-carbonium ion (*242*) produced by the cleavage of (*241*) can readily add back

[13] (i) for example, which has only one lone pair *antiperiplanar* to each C–O bond, is quite stable under conditions which lead to the rapid hydrolysis of acyclic or conformationally flexible orthoesters, like (*236*) [234].

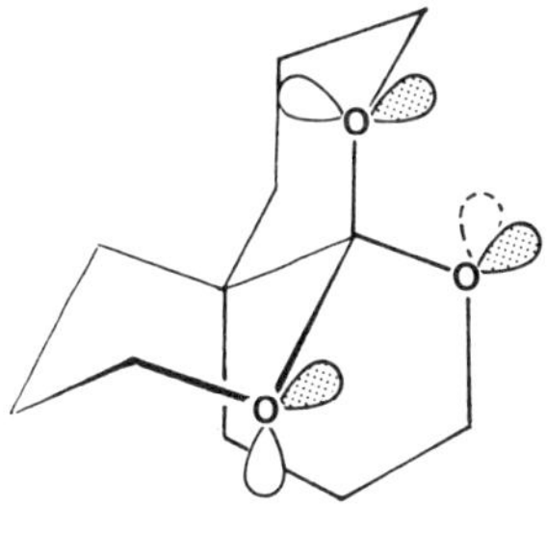

(i)

241 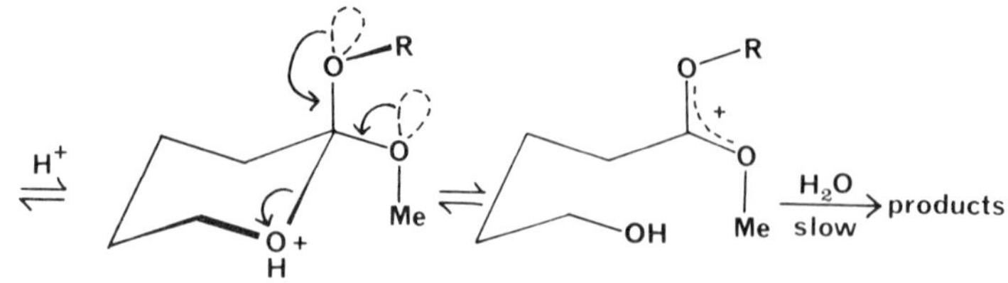

241a *242*

to regenerate (*241a*). In fact the hydration of (*242*), rather than its formation, is expected to be the rate determining step of the ring-cleavage reaction, because the alcohol OH group is not only a better nucleophile than external water, but its effective molarity will be several hundred times greater because its reaction is intramolecular. (The closest available comparison is the base-catalysed lactonisation of *243*, where the effective concentration of the OH group is 280M.) [214].

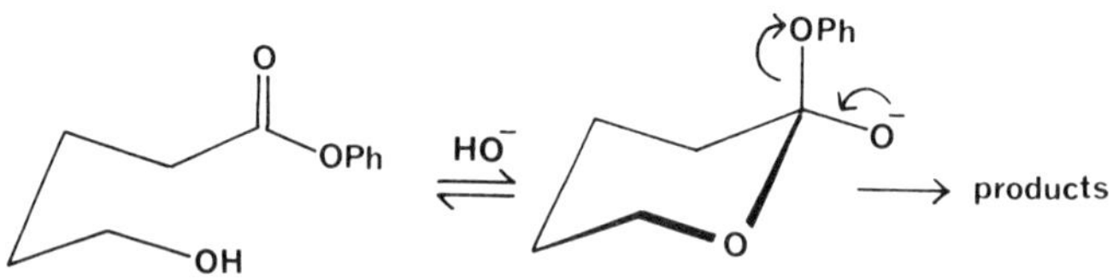

243

This factor manifests itself in terms of a reduced concentration of the intermediate (*242*), compared with that (*247*, below) from exocyclic C–O cleavage, and thus in a correspondingly reduced rate of hydrolysis by this route.

Of the three conformational isomers which remain (*244*–*246*), the second (*245*) is a high energy form because of the 1,3-diaxial-type interaction between the OR and OCH_3 bonds, and can be ignored.

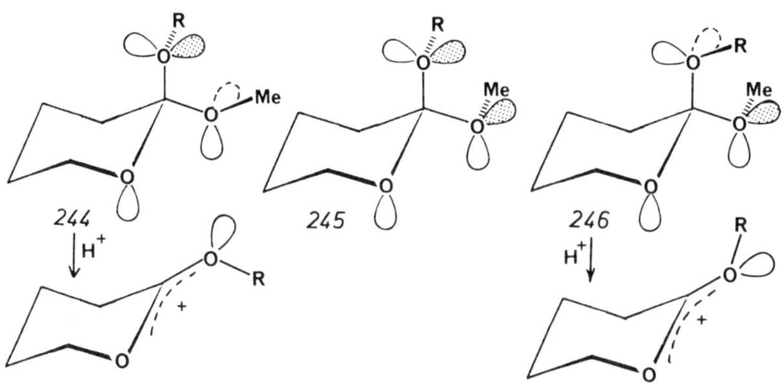

244 *245* *246*

247 *248*

Both (*244*) and (*246*) on the other hand, represent stable conformations, which can be cleaved with stereoelectronic control, with the loss of the axial leaving group as ROH; but the dioxocarbonium ions produced (*247* and *248*, respectively) have different conformations. Of these (*247*) is expected to be more stable (lone pair *antiperiplanar* to C–O$^{(+)}$ rather than C–C), and this difference in energy will be reflected to some extent in the transition states leading to (*247*) and (*248*). Thus (*244*), (with conformations A1, E3 about the axial and equatorial C–O bonds) is expected to be the reactive conformer of an orthoester of the general structure (*240*), for steric and stereoelectronic reasons, reinforced by a secondary kinetic anomeric effect: that is to say, an effect on the rate due to the orientation of lone pairs on the *product*, rather than the starting material, of the reaction.

a. Hemiorthoesters

The final steps in the hydrolysis of an orthoester involve the formation and cleavage of a hemiorthoester intermediate (*249*).

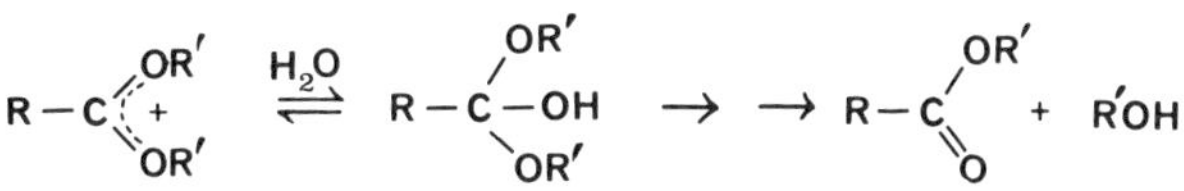

249

In principle an unsymmetrical acyclic hemiorthoester can break down to give four different products (*251–4*), and unless the leaving group capabilities of the two OR groups are very different we may once again expect stereoelectronic factors to be involved in determining the distribution of products.

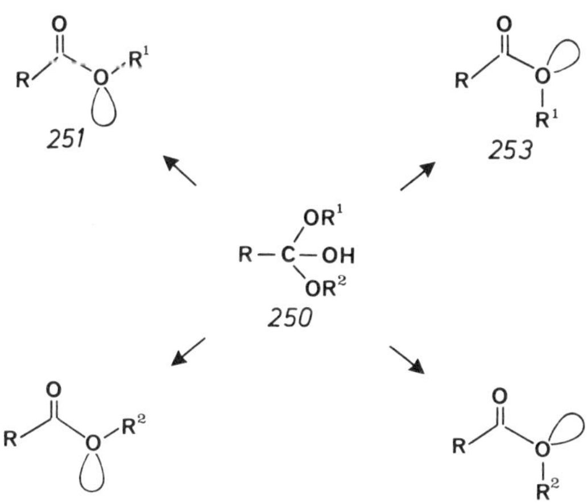

251 *253* *250* *252* *254*

Any secondary kinetic anomeric effect will favour the products with the Z conformation (*251, 252*) because this is the most stable conformation of the ester group, with the non-bonding electron pair on the oxygen of the OR group *antiperiplanar* to the C=O bond (see page 73).[14] Where there is a choice, therefore, we may expect a hydroxyester to be produced in preference to a lactone.

Experimentally a clear preference is observed for acyclic ester products. Orthoesters derived from δ- valerolactone and γ-butyrolactone are hydrolysed specifically to the hydroxyesters, rather than to the lactones [387].

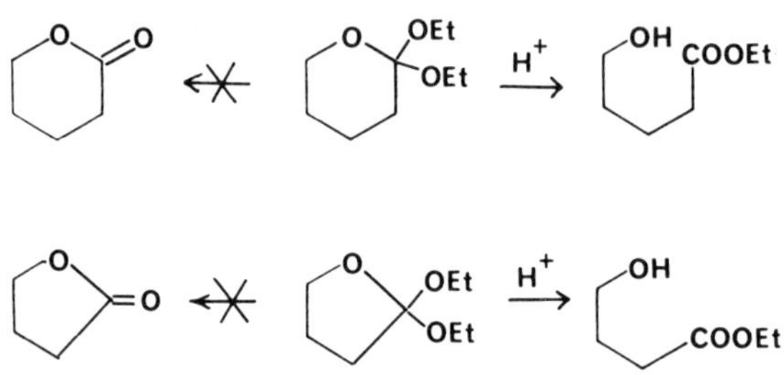

And a quantitative analysis of the pH-dependence of the product ratio from the hydrolysis of the phthalide orthoester (*255*) shows two pH-independent regions, corresponding to the acid and base-catalysed breakdown of the hemi-orthoester (*256*) [307b]. In acid 24% of lactone is produced, but above pH 3.5 the very rapid hydroxide catalysed breakdown of (*256*) gives only 7% of lactone, and some 93% hydroxyester.

[14] A related secondary effect is probably involved in the reaction of iodide with dioxocarbonium ions such as (*i*) [307a]. Although the reaction must be concerted displacement of oxygen, and reaction at the methyl centre would therefore be expected to be substantially faster than at the ring methylene group, the cleavage of (*i*) produces more iodoester (*iii*) than lactone product. This can be explained in terms of the greater stability of the ester group (in the Z-conformation) of (*iii*), which is thus a better leaving group than the lactone.

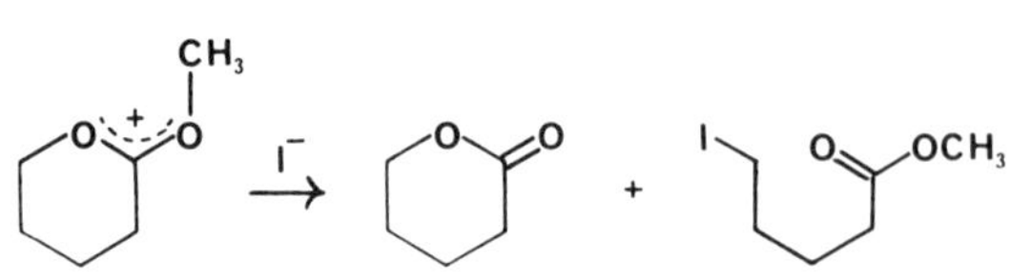

(i) (ii) (iii)

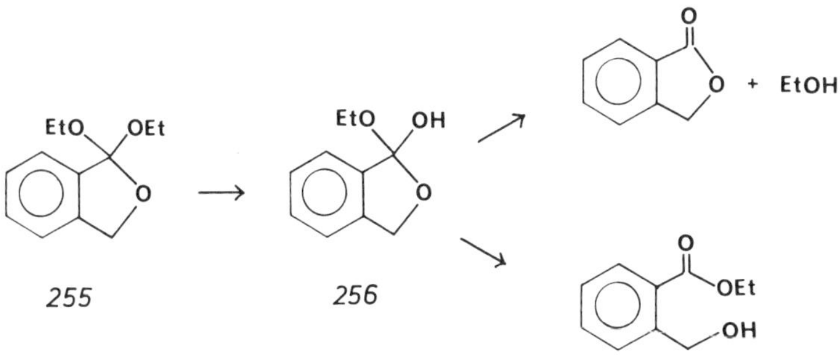

255 256

The main reason for this general preference for ring opening is not simply the secondary kinetic anomeric effect discussed above: a primary stereo-electronic effect may contribute also. In the hydrolysis of a cyclic orthoester (*257*) the initial step (page 102) is the loss of the axial OR′ group, as R′OH.

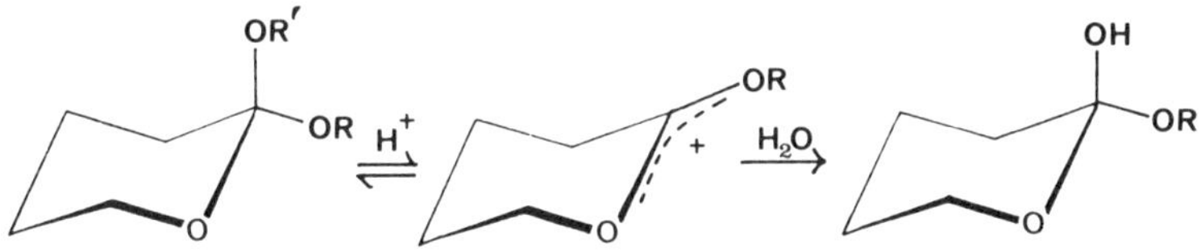

257 258 259

The hydration of the dioxocarbonium ion produced (*258*) then leads specifically to the axial hemiorthoester (*259*). (This is expected on grounds of microscopic reversibility, and from experiments (page 102) involving the addition of RO^- to the dioxocarbonium ion (*239*) – a conformationally more rigid version of (*258*) – which have been shown to produce the orthoester (*236*) with the OR group in the axial position.) And since it is a clear prediction of the stereo-electronic theory that an equatorial leaving group is never lost directly, the hydrolysis of (*259*) can only lead to the hydroxyester. (Thus of the three C–O bonds at the orthoester centre of (*257*), the axial, C–OR′ bond breaks the most, and the equatorial C–OR bond, the least easily.)

This analysis is confirmed and extended by the results of experiments in which hemiorthoesters are generated under more controlled conditions. The simplest technique is to synthesise on orthoester with a protecting group (R) on one oxygen, which can be removed by a reaction involving cleavage of the O–R bond. This approach has been used with notable success by Deslongchamps and his coworkers [308], who prepared the axial and equatorial isomers of the enol ethers (*261*) and (*262*). Oxidative cleavage of the enol ether by potassium permanganate in buffer at pH 13 then generated the axial and equatorial hemiorthoesters (*263*) and (*264*). Under the conditions these broke down immediately, and the initial product from both isomers was the hydroxyester (*237*).

This is exactly as predicted by the stereoelectronic theory for the axial OH compound (*263*), as discussed above. The equatorial isomer (*264*), on the other hand, has a favoured conformation with two oxygen lone pairs *antiperiplanar* to both the OCH_3 and the ring C–O bond, so that either could be cleaved with stereoelectronic control. The formation of the hydroxyester (*237*) rather than the lactone is therefore presumably controlled by the secondary kinetic anomeric effect (or secondary stereoelectronic effect [308]) discussed above.

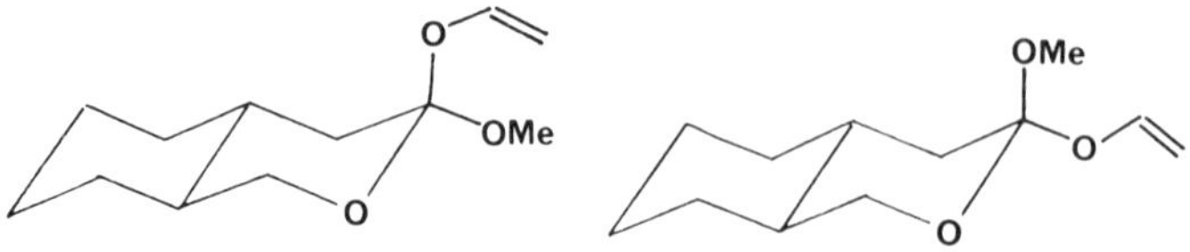

261 *262*

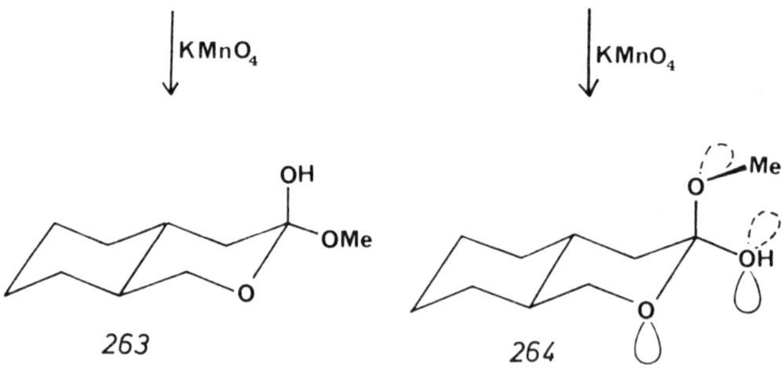

263 *264*

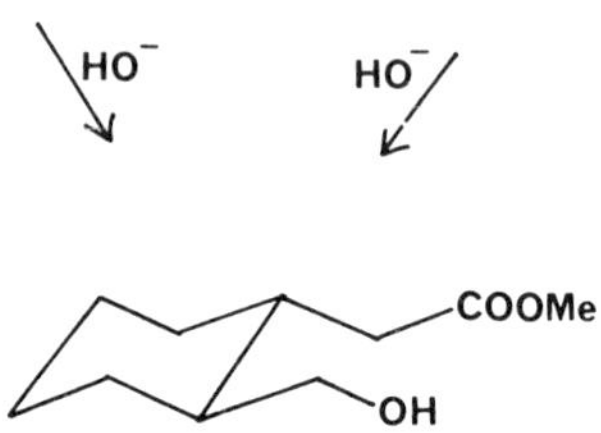

237

An alternative approach is to generate the hydroxyorthoester by the addition of water to the parent dioxocarbonium ion, as in the hydrolysis of orthoesters, but under conditions in which it breaks down more slowly than it is formed. This technique has been used by Capon et al. [309] for the preparation of simple hydroxyorthoesters like (*265*) and (*266*):

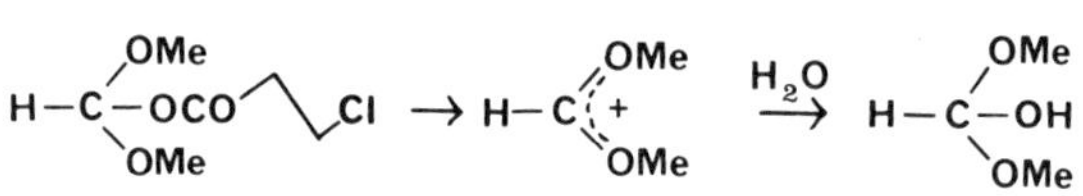

265

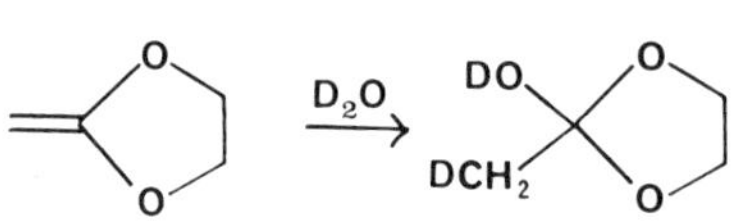

266

and by Kresge and McClelland and their co-workers, who observed that the breakdown of the hemiorthoester may be the rate determining step in the hydrolysis of certain orthoesters at low pH [310]. These kinetic studies show that simple hemiorthoesters, though very reactive species, can have significant lifetimes, of the order of a second under suitable conditions [307a].

More commonly the preformed dioxocarbonium ion is used. King and Albutt [311], in a very early demonstration of stereoelectronic control of hydroxyorthoester cleavage, generated the intermediate (*267*) from (*268*), and showed that it breaks down to give the axial ester (*269*) almost exclusively. (This paper contains a very clear statement of the basic tenets of the stereoelectronic theory, as applied to the reactions of esters and orthoesters.) [311] The same isomer is produced, probably exclusively, by the ozonolysis of the corresponding acetal (*270*) [304].

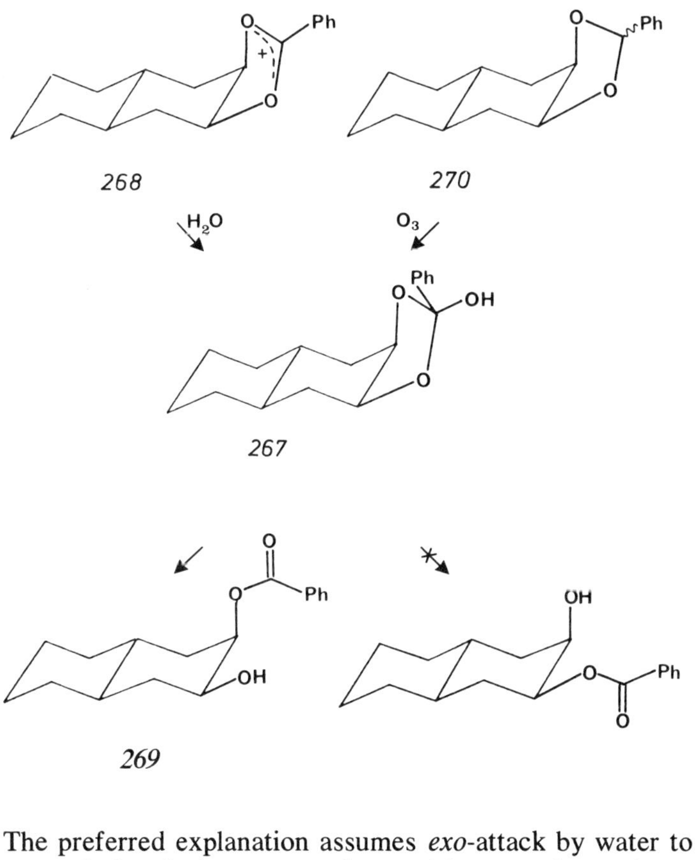

The preferred explanation assumes *exo*-attack by water to give (*267*), which can only break down to products with stereoelectronic control by assuming twisted conformations (*271, 272*). (Because n-σ^*_{C-O} overlap is impossible for the lone pairs of the ring oxygens of a 5-membered cyclic acetal or orthoester when the O–C–O system in the ring is planar. See page 91.

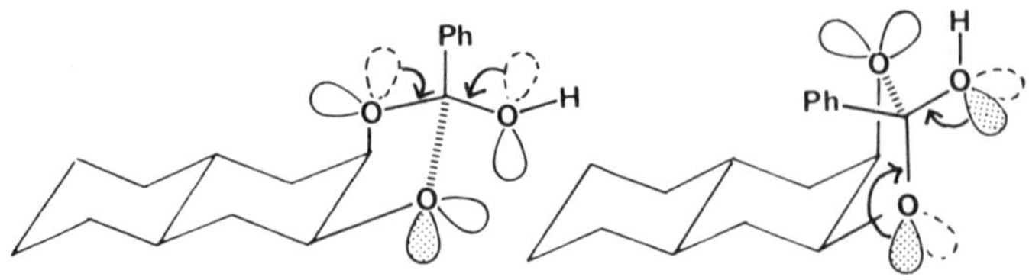

271 272

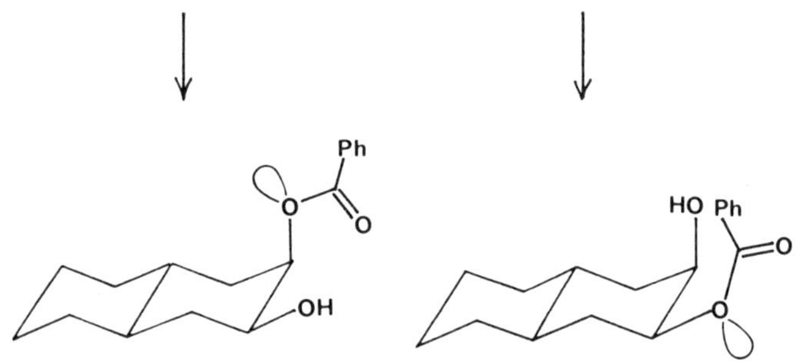

269 273

A twist which takes the benzene ring away from the decalin (to give *271*) is preferred to the one in the direction of (*272*), which would introduce severe non-bonded interactions with the axial hydrogens, so that the pathway to the axial ester (*269*) is favoured. Note also that reaction, via these two conformations at least, would produce the equatorial ester in the less stable E conformation (*273*), so that a secondary kinetic anomeric effect also favours the formation of the axial ester.

The same technique, of adding the nucleophile to the preformed dioxocarbonium ion, has been used to generate the SH analogues (e.g. *274*) of hemiorthoesters [312].

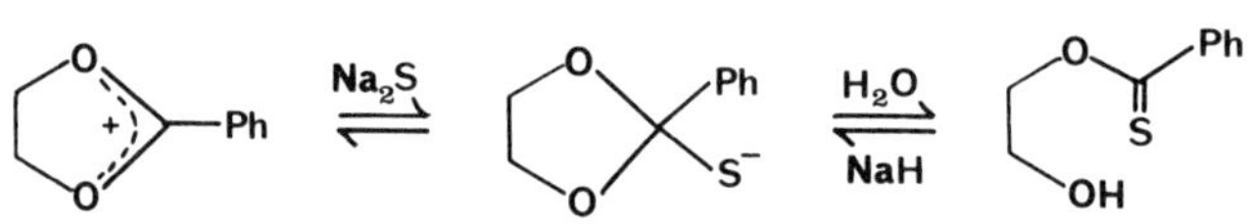

274 275

In the presence of a proton source these intermediates break down to give hydroxyalkyl thionesters (*275*), and the selectivity of bond cleavage in the six-membered cyclic systems (*276–278*) is much the same as that observed for the corresponding all-oxygen series, suggesting that stereoelectronic effects are qualitatively similar at oxygen and sulphur [313].
Compound (*276*) for example, in which either C–O bond can be cleaved with stereoelectronic control, breaks down to give comparable amounts (about

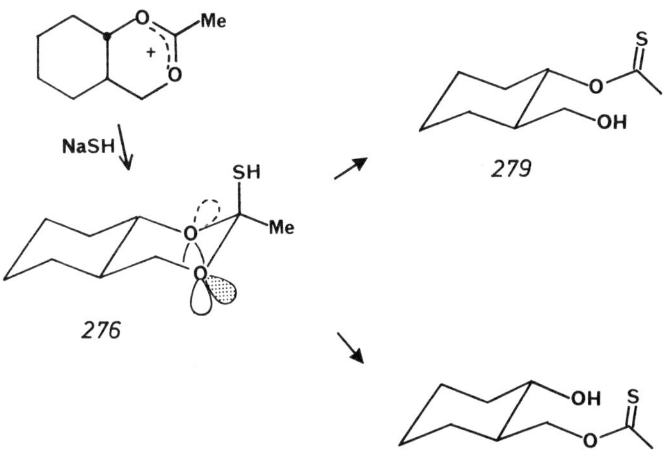

276 279 280

2 : 3) of (*279*) and (*280*). On the other hand, the addition of NaSH to (*281*) gave no hydroxythionolactone (*282*), but what was apparently a mixture of (*277*) and its *cis*-isomer (*283*) [313].

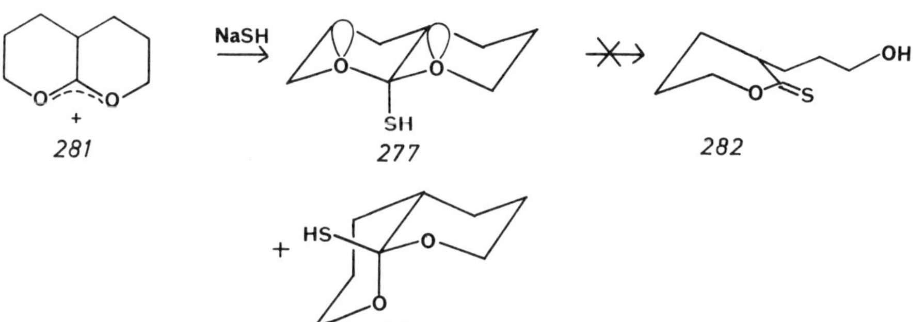

281 277 282 283

It is expected that (*277*), which has only equatorial leaving groups, should not break down to give (*282*); and that it should eliminate SH to regenerate (*281*) and hence eventually give a thermodynamic mixture of *cis* and *trans* isomers (*277* and *283*). But it is surprising, if all this is true, that no (*282*) is produced eventually, because it could be formed with stereoelectronic control from (*283*).

The third intermediate (*278*) on the other hand, behaves just as predicted, breaking down to give the hydroxy thioester [313], rather than the thiolactone, as do the corresponding OH compounds (page 108).

Thus there is rather convincing evidence that the breakdown of orthoesters and their partially hydrolysed derivatives is subject to stereoelectronic control. If there is no available conformation in which one or more of the three central

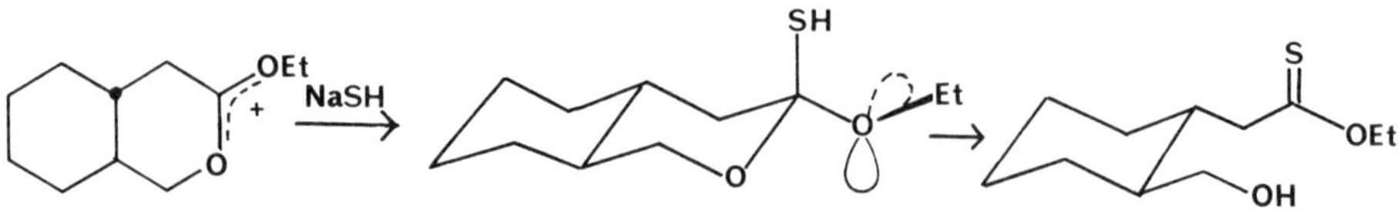

278 284

C−O bonds is *antiperiplanar* to lone pairs on the other two oxygen atoms the compound is kinetically stable (see footnote to page 103). If such a conformation is available then that bond (or bonds) is potentially labile; though the compound may of course still be thermodynamically stable in the tetrahedral form, and a number of examples are known of what have been called 'stable tetrahedral intermediates'. Several of these are derived from the highly reactive trifluoroacetyl group, and must be thermodynamically stable with respect to elimination (*285–7*) [314–316].

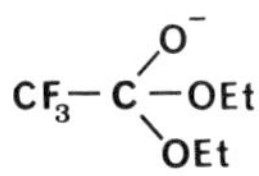

285

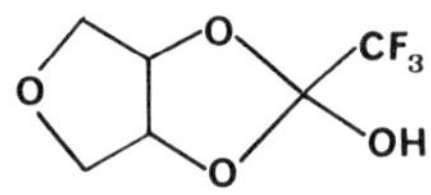

286

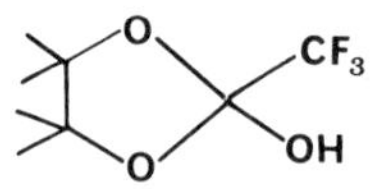

287

The most celebrated example, which is stable even as the anion (*288*) is the fish poison tetrodotoxin [317].

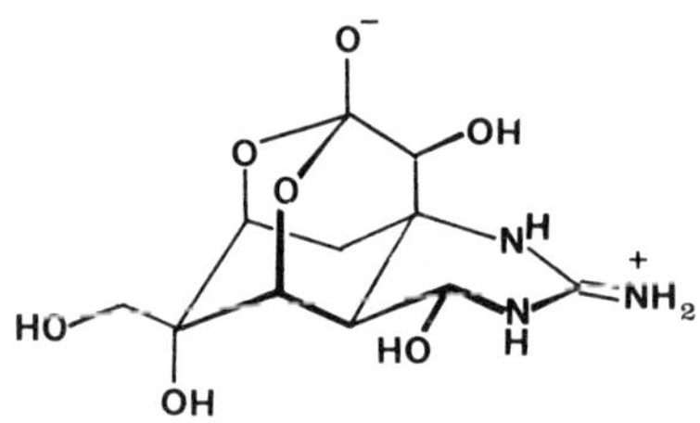

288

4. Carbonyl-Oxygen Exchange Reactions of Esters and Amides

Hemiorthoesters with *two* OH groups are the final tetrahedral intermediates in the hydrolysis of orthoesters, but are more familiar as the intermediates involved in the $A_{AC}2$ and $B_{AC}2$ mechanisms of ester hydrolysis [318].
The intermediate (*289*), formed by the addition of OH^- to an ester in the usual Z conformation, has two lone pairs on each oxygen atom *antiperiplanar* to each C−O bond (if we assume that the nucleophilic OH takes up an appropriate conformation or, more generally, that (*289*) is in rapid prototropic equilibrium,

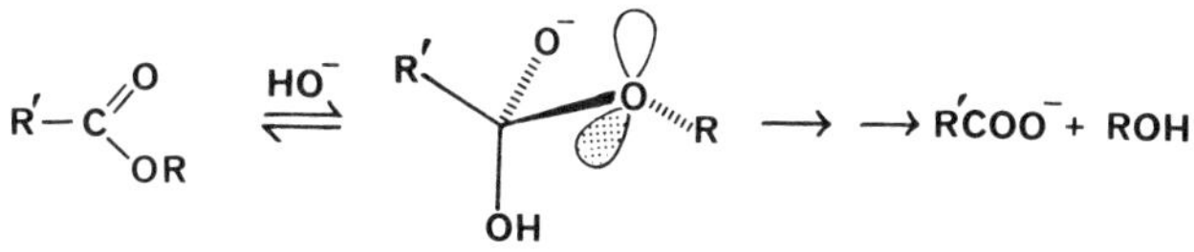

289

so that lone pairs and OH bonds may be regarded as equivalent). So there are no stereoelectronic barriers to the breakdown of (*289*).

The situation is different when the ester is in the E conformation (*290*). Here it is the OR bond of the ester which is *antiperiplanar* to the C=O bond, and it remains *antiperiplanar* to the developing C−O⁻ bond of the tetrahedral intermediate (*291*). The hydrolysis pathway will not be affected, but there is clearly a potential stereoelectronic barrier to the loss of what was the carbonyl oxygen atom, which can be monitored if it is isotopically labelled.

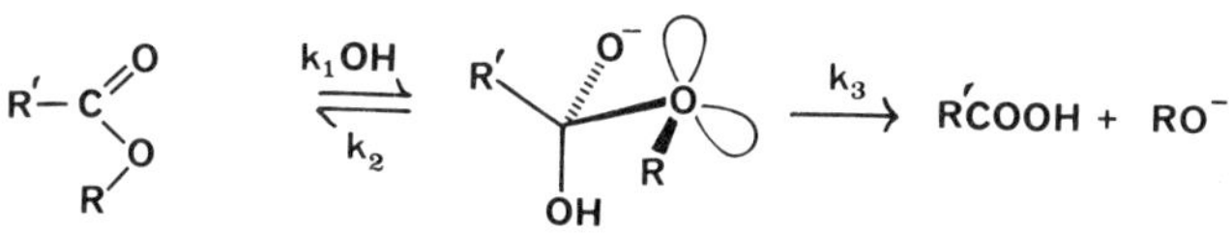

290

291

Thus in the case of a lactone (*292*), the necessary proton transfers are not sufficient to lead to ^{18}O-exchange, even if this is electronically favourable ($k_3 < k_2$), because this would mean losing OH^- without stereoelectronic control

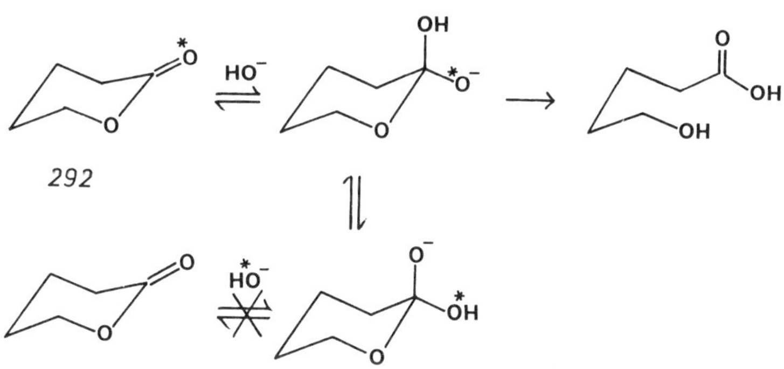

292

293

(e.g. from the equatorial position of (*293*)) [319]. These predictions are in agreement with the known experimental facts. Acyclic esters show ^{18}O exchange in alkali which occurs simultaneously, and often at a rate comparable with hydrolysis [318]; but no exchange reaction is observed with γ-butyrolactone, phthalide [320, 321], δ-valerolactone [322] or (*294*) [319].

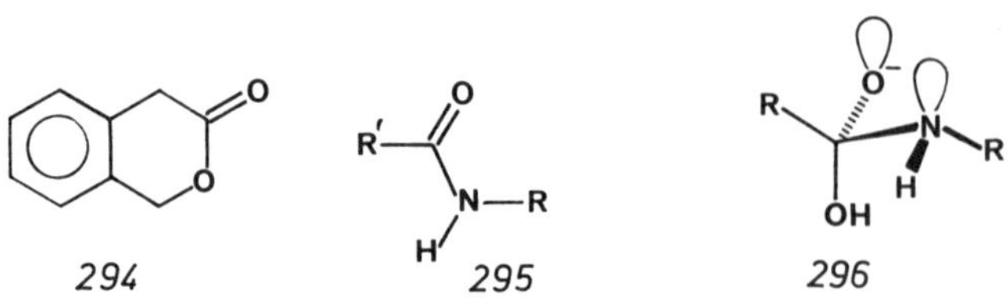

With amides the picture is more complicated. Primary and secondary amides (which exist in the Z conformation (*295*)) will react with hydroxide to give initially (*296*); which, like the corresponding adduct (*289*) from an ester, has no stereoelectronic barrier to the cleavage of any one of the three central carbon-heteroatom bonds (N–H being regarded as equivalent to a lone pair). Consequently the relative rates of hydrolysis and carbonyl oxygen exchange should be governed by simple electronic factors. And since RNH^- is a poorer leaving group than HO^- extensive exchange is expected; and observed, for primary and secondary amides [320].

Tertiary amides, and all lactams, on the other hand, have a N–C bond in the position occupied by the N–H bond of (*295*), so that the addition of hydroxide leads initially to a tetrahedral intermediate (*297*) which, even if the necessary proton transfers take place (to give *298*), can only break down to give products, or to regenerate starting materials.

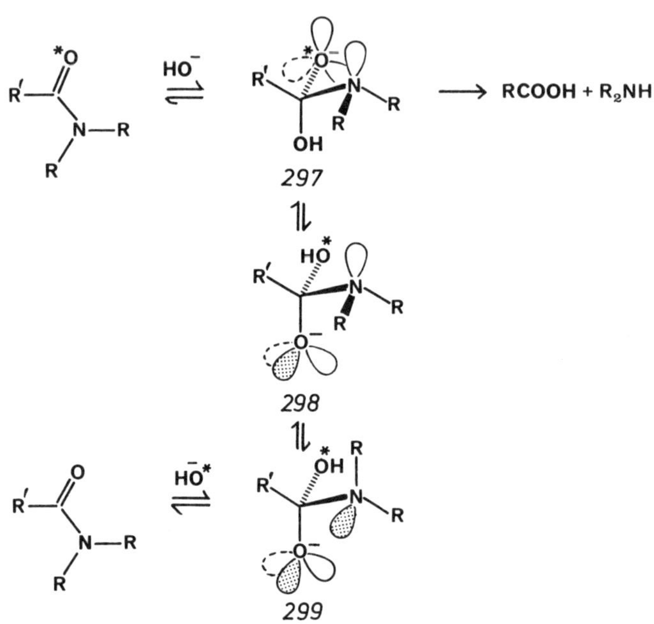

If the stereoelectronic factors which govern participation by the nitrogen lone pairs are as strict as seems to be the case for oxygen, exchange will require a rotation about the C–N bond of (*298*) to give a conformation (*299*) with lone pairs on both O^- and N *antiperiplanar* to the C−*OH bond. (Specifically, this requires either a rotation of 120°, or nitrogen inversion followed by a rotation of 60°) [319].

In fact oxygen exchange does accompany the hydrolysis of certain tertiary amides. The factors governing the rate of the exchange reaction have been studied in some detail, and in some cases at least there is evidence to relate the reaction with the ease of rotation about the C−N bond.

Only two lactams have been examined, and neither piperidone nor N-methylpiperidone shows significant ^{18}O-exchange on hydrolysis in KOH at room temperature [319]. Furthermore, the work of Bunton et al. [323] on benzamide and its N-methyl derivatives shows that the extensive exchange observed with $PhCONH_2$ and PhCONHMe does not occur during the alkaline hydrolysis of $PhCONMe_2$.

On the other hand, the hydrolysis of N,N-dialkylformamides is accompanied by oxygen-exchange, which is generally faster than hydrolysis at elevated temperatures. Table 22 shows some of the data for the relative rates of hydrolysis and exchange (thus correcting for absolute reactivity) of a series of tertiary amides studied by Deslongchamps and his group [324, 325]. The ratios show a strong temperature dependence: the exchange reaction has a significantly higher enthalpy of activation, and is thus relatively more important at higher temperatures [325]. Conversely, hydrolysis shows a very large negative entropy of activation (of the order of −40 e.u., compared with −(20–25) e.u. for exchange) [325], which is ascribed to the need for a molecule of solvent in the transition state to protonate the developing R_2N^- group, which would not readily depart otherwise (*300*).

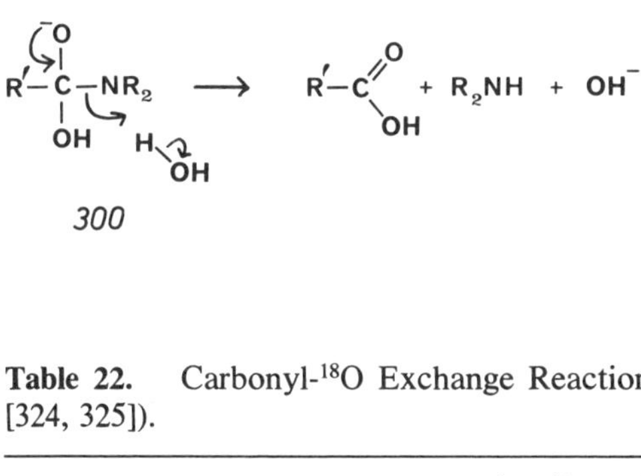

300

Table 22. Carbonyl-^{18}O Exchange Reactions of Tertiary Amides (Data from Refs. [324, 325]).

	k_{hyd}/k_{exch} (0.15 M KOH at 90°) for		
R_1, R_2	$HCONR_1R_2$	$MeCONR_1R_2$	$EtCONR_1R_2$
$PhCH_2$, Me	0.87	2.5	4.9
Et, Et	–	2.0	–
i-Pr, *i*-Pr	0.35	–	–
Ph, Me	0.89	1.06	0.78
2,6-Dimethyl-phenyl, Me (*301*)	0	–	–

(When this proton transfer is inhibited sterically (compound *301*, Table 22), exchange occurs in the absence of hydrolysis.)

The ratio k_{hyd}/k_{exch} is typically in the range 0.2–0.5 for the alkaline hydrolysis of primary and secondary amides and anilides [320], so there is little evidence for stereoelectronic control in the data of Table 22 for formamides or formanilides. On the other hand there is a significant increase in the ratio for N,N-dialkyl acetamides and propionamides [325]. This could result either from an increase in k_{hyd} or a decrease in k_{exch}, compared with the formamides: and both might be expected. On the one hand the non-bonded interactions in the intermediate (*300*), which build up with increasing size of the acyl substituent R′ (one result is the lower overall reactivity of propionamides compared to acetamides compared with formamides), are relieved more directly by breaking the C−N rather than a C−O bond. On the other hand, the same steric congestion might be expected to slow rotation about the C−N bond.

Deslongchamps at al. have analysed the activation parameters for the alkaline hydrolysis of the three N-benzyl N-methyl amides of Table 22 [325] in terms of the simplified Scheme 1.

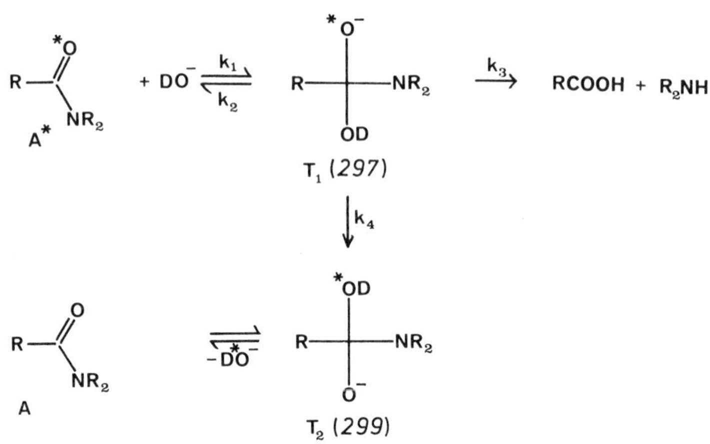

Scheme 1

The tetrahedral intermediates T_1 and T_2, which can lose DO^- or D^*O^- respectively, correspond to (*297*) and (*299*), above (A is of course also eventually hydrolysed). Using estimates for the equilibrium constants k_1/k_2, based on Guthrie's work [326], and *assuming* that the breakdown of the tetrahedral intermediates to starting material is much faster than their interconversion ($k_2 \gg k_4$), values could be calculated for k_3 and k_4. These suggest that the product-forming step (k_3) is actually *slower* for the more hindered intermediates (T_1 = (*297*), R′ = Me, Et) but that the rate of rotation about the C−N bond (k_4) is slowed still more by the larger acyl groups. The estimated barriers to the interconversion (*297* ⇌ *299*) are 5.8 and 8.2 kcal mol^{-1} for R′ = H and R′ =

Et respectively [325], clearly larger than the barriers expected for simple proton transfers, and not unreasonable for barriers to rotation about heavily substituted single bonds.

These results are consistent with a stereoelectronic barrier to the loss of the original carbonyl oxygen atom from the first-formed tetrahedral intermediate in the hydrolysis of tertiary amides, other than formamides. (Anilides also appear to be exceptions, as far as can be judged from the data of Table 22.)

In view of the assumptions involved, however, they do not constitute a proof of the existence of such a barrier. Fortunately more evidence is available about the behaviour of the corresponding intermediates formed in the hydrolysis of imidate esters, where the two oxygen atoms are distinguished by alkylation, rather than by isotopic labelling.

5. The Hydrolysis of Imidate Esters: Stereoelectronic Effects at Oxygen and Nitrogen

The hydrolysis of an imidate ester (*302*) can lead, depending on the conditions, to ester plus amine or amide plus alcohol products.

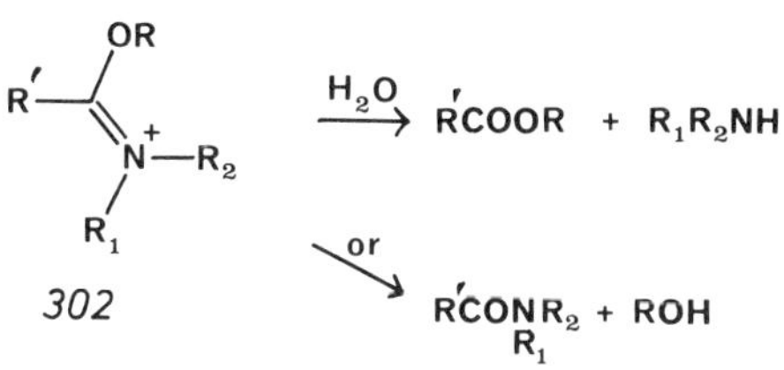

302

One might expect the amine to be the favoured product in acid, where the tetrahedral intermediate T_+ (Scheme 2) is protonated, but not in base, where the leaving group from T_- would be formally the anion R_2N^-. In practice these remain useful guidelines to product ratios, but they are far from being generally applicable, mainly because they take no account of stereoelectronic control. In particular C−N cleavage is commonly faster than loss of RO^-, even at high pH.

The generally accepted mechanism for the hydrolysis of an imidate ester [327] is shown as Scheme 2. Not all the tetrahedral intermediates will exist for a given compound under all conditions, nor are they necessarily in protonic equilibrium. But the rate determining step of the hydrolysis reaction is usually the addition of hydroxide or water to give the tetrahedral intermediate T_0 or T_+.

At low pH hydrolysis goes through the zwitterionic intermediate $T_\pm$. The breakdown of $T_\pm$ can be a very fast reaction – clearly faster than rotation about single bonds in some cases [327a] – but it can only break down in one direction (loss of RO^- would generate an N-protonated amide, so there is insufficient driving force for the elimination of what is in any case the poorer leaving group). Consequently ester and amine are the only products under

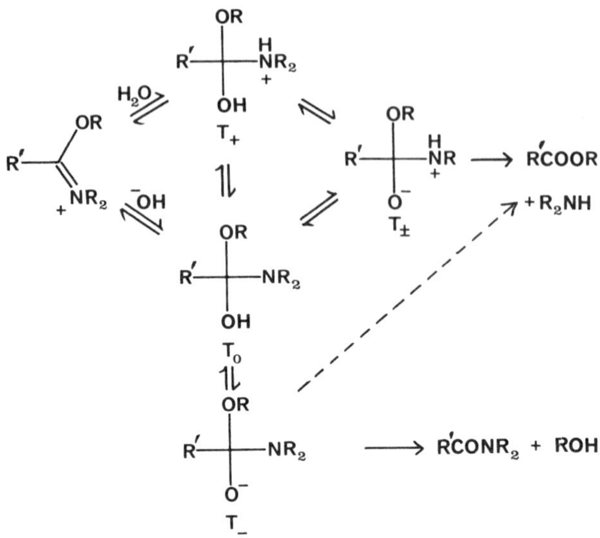

Scheme 2

neutral or acidic conditions. At high pH, however, hydrolysis can occur *via* T_-, corresponding to the tetrahedral intermediates involved in the base catalysed oxygen exchange reactions (Scheme 1, page 116). Clearly

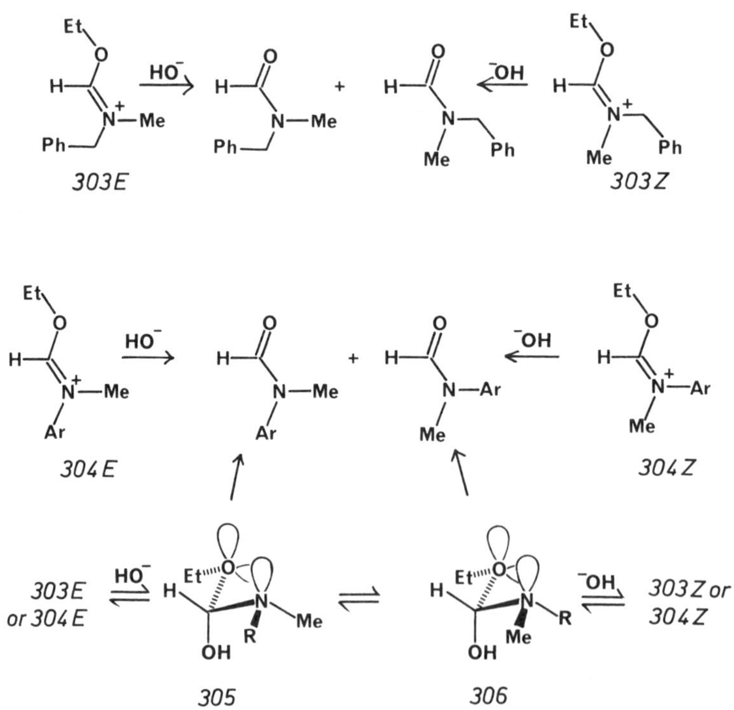

stereoelectronic effects at nitrogen will be expected to be similar in the two systems, though the further complication of conformational isomerism about the C–OR bond arises for the intermediates from imidate esters. We will consider first stereoelectronic effects at nitrogen.

Several experiments suggest that rotation about the C–N bond of tetrahedral intermediates derived from unsymmetrical imidate esters can be rapid. The hydrolysis of two formic acid derivatives (*303* and *304*, Ar = 2,6-dimethylphenyl) give mostly (75 %, 100 %) amide product under strongly basic conditions [328]. In both cases the individual E and Z isomers give the same mixture of E and Z amides, showing that the tetrahedral intermediates (*305* and *306*) must be readily interconvertible under the conditions.

A similar experiment of Caswell and Schmir [327] gave a similar result. Since the product ratios from the N,N-dimethyl and N,N-di-isopropyl (*307*) imidate esters are very different (*308*, R = Me, gives 81 % of amine plus ester [330] whereas (*308*, R = i-Pr) gives only 6 % [327]), it might reasonably be expected that the E and Z-isomers of the N-methyl, N-iso-propyl compound (*308*) would give different product ratios if the stereochemical integrity of the dialkylamino-group is retained.

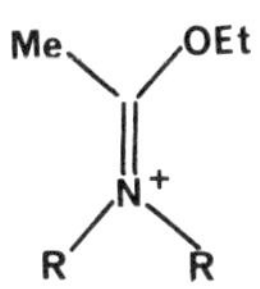

307

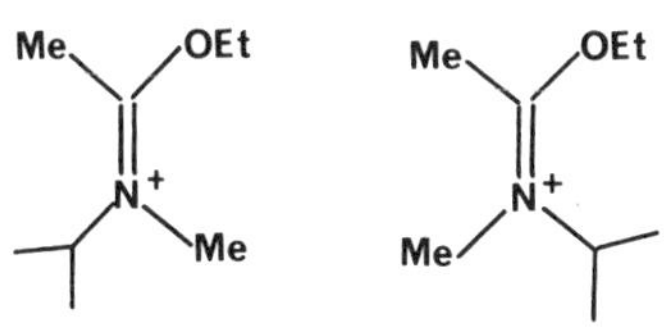

E *308* Z

In fact above pH 12 both the pure E isomer and a 1 : 1 mixture of E and Z isomers of (*308*) gave the same product mixture, containing 65 % amine plus ester. The result is slightly complicated by a slow interconversion of E and Z (*308*) by way of the enamine, but this is not fast enough to account for the identical product ratios obtained from the two isomers. Clearly even in the case of the acetic acid derivative (*308*) there is time for the interconversion of tetrahedral intermediates (corresponding to (*305*) and (*306*)) before they break down.

Only when the lifetime of the tetrahedral intermediate is reduced still further is there any indication that cleavage can become faster than rotation about a single C–N bond. The intermediate (e.g. *309*) derived from an amidine breaks down via the zwitterion, in which C–N cleavage is assisted by lone pairs on the second nitrogen atom as well as on O^-. There is no evidence that restricted rotation about a C–NAr bond affects the product distribution from an amidine [329a], but the product of alkaline hydrolysis of N,N′-dimethylformamidine (which exists predominantly (95 %) in the E, Z conformation shown) is E-N-methylformamide, rather than the thermodynamically favoured Z-isomer [329b].

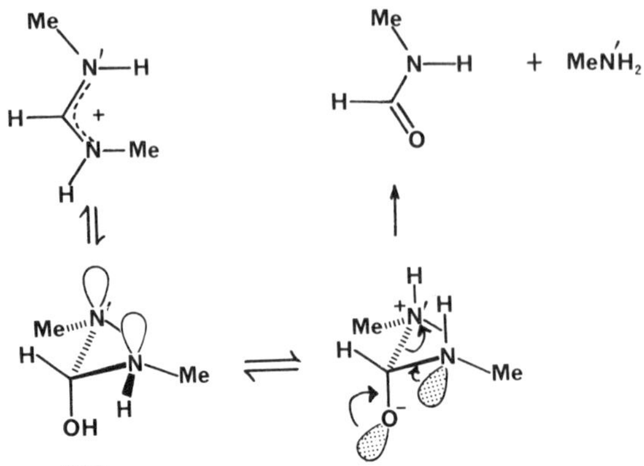

309

This is readily explained in terms of the theory of stereoelectronic control, because the initially formed tetrahedral intermediate (*309*) has lone pairs on both O and N *antiperiplanar* to the C−N′ bond (protons and lone pairs being considered equivalent, as discussed above). Cleavage of this bond via $T_{\pm}$ generates the E-amide directly, whereas cleavage of the other C−N bond, which would lead to the more stable Z-amide, requires a rotation about C−N′.

This, and the ^{18}O-exchange results with heavily substituted tertiary amides, suggest that stereoelectronic effects involving the lone pair electrons of nitrogen can be important in determining the direction of cleavage of tetrahedral intermediates, but only in specially favourable cases. More convincing evidence comes from extensive studies on product ratios from the alkaline hydrolysis of imidate esters.

The course of the basic hydrolysis of imidate esters clearly depends on the geometry about the C−OR bond. Table 23 summarises measured product ratios for the reaction of imidate esters without special complicating features. The cyclic esters at the top of the list, which are constrained in the Z conformation about the C−OR bond, give the ester plus amine product exclusively. On the other hand the lactone analogues at the end of the Table, which are fixed in the E conformation, give mixtures of amine plus ester (lactone) and amide plus alcohol. The open chain esters, and those cyclic compounds without a fixed conformation about the C−OR bond, fall into the same pattern. Formic and acetic acid derivatives are largely if not exclusively in the E conformation, as shown, for example, by a nuclear Overhauser effect study on ester (*310*).

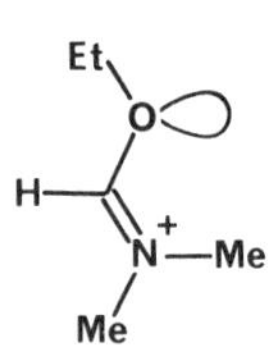

310

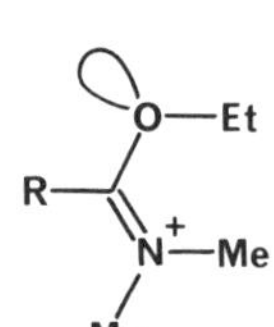

311

Table 23. Product Ratios from the Hydrolysis of Imidate Esters Under Basic Conditions

Compound		*Products* (%)		
		Ester + Amine	*Amide + Alcohol*	*Ref.*
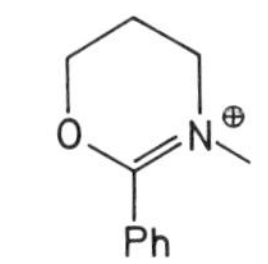		100	0	[242]
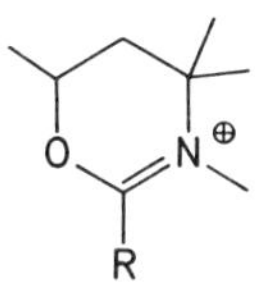	R = Me	100	0	[330]
	R = Ph	100	0	[330]
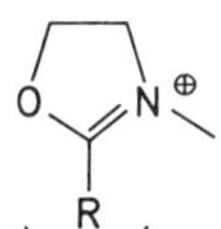	R = Me	100	0	[330]
	R = Ph	100	0	[330]
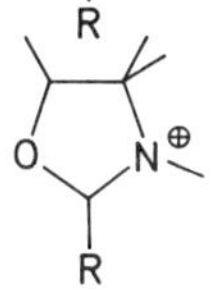	R = Me	100	0	[331]
	R = Et	100	0	[331]
	R = t-Bu	100	0	[331]
	R = Ph	100	0	[331]
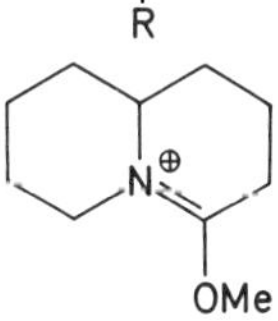		100	0	[330]
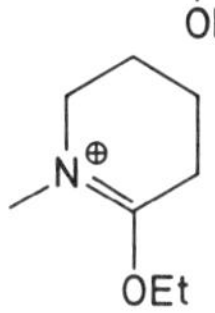		83	17	[330]
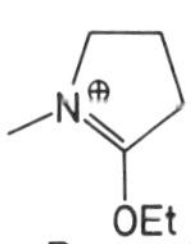		65	35	[330]
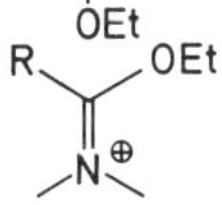	R = t-Bu	100	0	[330]
	R = Ph	100	0	[330]
	R = cyclohexyl	75	25	[330]
	R = Me	81	19	[330]
	R = H	50	50	[330]
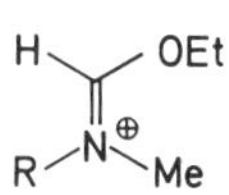	R = CH_2Ph	25	75	[328]
	R = Ph	50	50	[328]
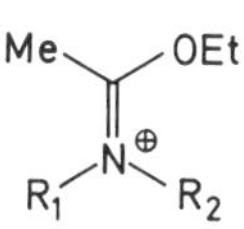	$R_1 = R_2 =$ Me	81	19	[242]
	$R_1 = R_2 =$ i-Pr	6	94	[329]
	$R_1 =$ Me, $R_2 =$ i-Pr	65	35	[327]
	$R_1 =$ i-Pr, $R_2 =$ Me	65	35	[327]

Table 23. (cont.)

Compound	*Products* (%)		
	Ester + Amine	*Amide + Alcohol*	*Ref.*
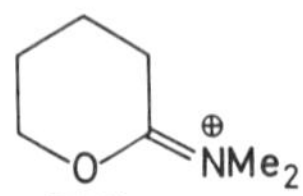	33	66	[330]
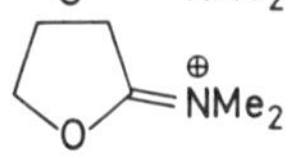	50	50	[330]

This avoids the 1,3-syn interaction with the N−Me group; but when the acyl substituent (R in *311*) is large enough the steric repulsion between it and the OEt group will be the dominant non-bonded interaction, and the Z conformation (*311*) may be expected to be preferred (e.g. for R = Ph or *t*-Bu). Thus the evidence is that imidate esters which are constrained in the Z conformation (*311*) are hydrolysed at high pH exclusively to the amine plus ester, whereas E compounds give a mixture of products [330].

This preference is only readily rationalised in terms of stereoelectronic control. The addition of hydroxide to a Z-imidate will generate a tetrahedral intermediate, T_- (*312*) which has two lone pairs *antiperiplanar* to the C−N

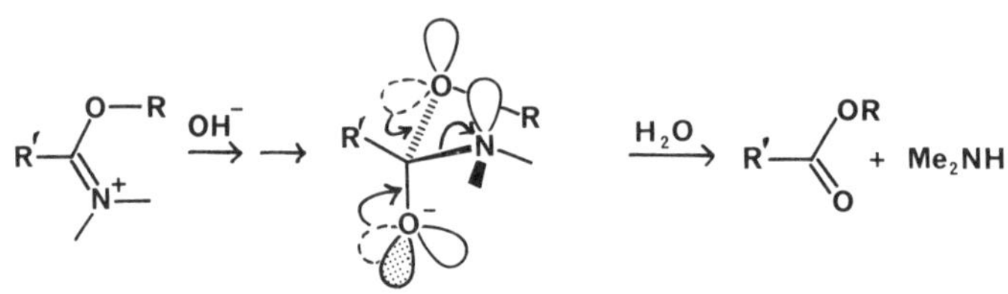

312

bond, but not to the C−OR bond. Consequently (*312*) can break down with stereoelectronic control to give ester plus amine products, without the need for conformational equilibration. The fact that no amide is produced would be explained if the loss of RO^- were intrinsically many times slower than C−N cleavage. But the evidence from the ^{18}O-exchange results is that the loss of HO^- from the similar intermediate (*298*, corresponding to *312*, R = H) is actually *faster* than C−N cleavage in the absence of a stereoelectronic barrier. The simplest explanation is that (*312*) breaks down to ester plus amine faster than rotation about the C−N bond can bring the nitrogen lone pair *antiperiplanar* to the C−OR bond. The breakdown of (*312*) is thus under stereoelectronic control.

The corresponding intermediate (*313*) from an E-imidate, on the other hand, has only one lone pair *antiperiplanar* to either the C−N or the C−OR bond. So neither can break with stereoelectronic control unless there is a rotation about the other. Compared with (*312*), which can break down to products from the conformation in which it is formed, (*313*) is thus expected to be significantly longer-lived.

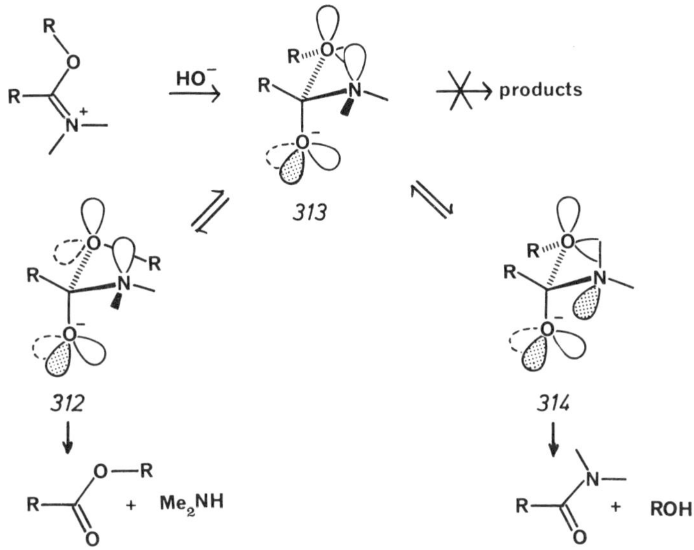

Rotation about the C–OR bond will produce (apart from a third rotamer which would give the ester in the unfavourable E-conformation) the same intermediate (*312*) which is involved in the hydrolysis of a Z-imidate, and gives only ester plus amine products. (This conformation (*312*) will be least favourable when the N-alkyl substituents, which are in a 1,3-*syn* relationship to the O–R bond, are large, as in the N,N-di-isopropyl compound (*308*) above; which gives significantly less amine plus ester than related E-imidates with smaller N-alkyl groups.)

Rotation about the C–N bond leads to (*314*), which can break down with stereoelectronic control to amide plus alcohol. Since comparable amounts of ester plus amine and amide plus alcohol are produced from the alkaline hydrolysis of E-imidates (Table 23), reaction presumably occurs to comparable extents via (*312*) and (*314*), as expected if the barriers to rotation about the C–OR and C–N bonds are similar. This is not readily distinguished in practice from the alternative situation where they are negligible, and stereoelectronic barriers are absent: since the product ratios from the basic hydrolysis of E-imidates are qualitatively similar to those expected from the ^{18}O-exchange results with primary and secondary amides, which show that the loss of HO^- from T_- is only a few times faster than C–N cleavage in the absence of stereoelectronic barriers. So the most convincing evidence that stereoelectronic effects at nitrogen can control the breakdown of tetrahedral intermediates remains the exclusive C–N cleavage observed in the basic hydrolysis of Z-imidates.

The large and complex body of results for the alkaline hydrolysis of imidate esters, and for the base-catalysed oxygen-exchange reactions of amides, can thus be rationalised in terms of the stereoelectronic theory: cleavage of any C–O or C–N bond proceeds easily only if there are lone pairs on both remaining O or N atoms *antiperiplanar* to it. The barrier to rotation about the C–N bond is generally smaller than that for cleavage of C–O, so that stereoelectronic control of C–O cleavage is only apparent in cases where rotation is hindered. The (presumably comparable) barrier to rotation about the C–OR bond of imidate esters similarly can be higher than that for C–N cleavage in some short-lived intermediates, and is responsible for the characteristic patterns of base hydrolysis products of the E and Z-esters.

Since none of these barriers involves more than a few kcal/mol the differences between them are quite small, and can be magnified or reversed by the special structural features of particular molecules. Thus C–N cleavage is inhibited almost completely, even in acid, for derivatives of 2,6-dimethylaniline [324, 328, 332, 333], which cannot easily be protonated on nitrogen. In cyclic compounds rotation about C–O or C–N bonds is restricted, and the stereoelectronic barriers consequently increased, accounting, for example, for the lack of ^{18}O-exchange in the hydrolysis of lactams [319] (page 115), and for the exclusive C–O cleavage observed in the base hydrolysis of (*315*).

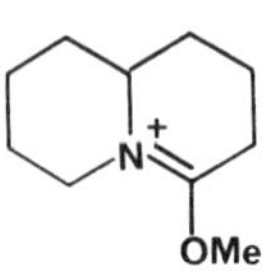

315

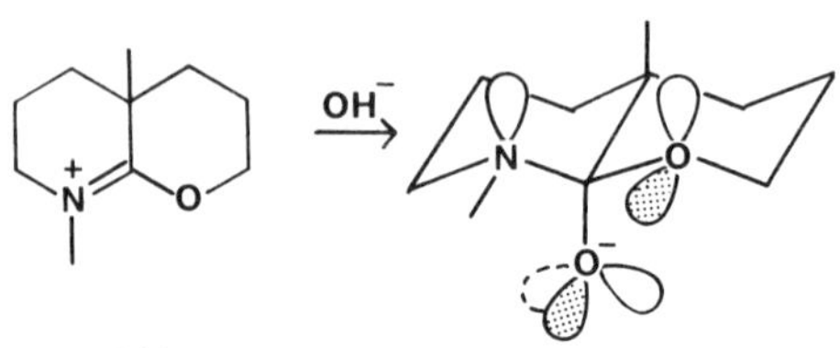

316 317

Finally, the stereoelectronic barriers are not insurmountable. Imidate ester (*316*) should give rise initially to the tetrahedral intermediate (*317*), which can neither break down to products with stereoelectronic control, nor find its way to a low energy conformation which can [330]. Nevertheless (*316*) is hydrolysed in base to the amide alcohol (*318*), and in acid to the ester amine (*319*), (which is rapidly transformed to (*318*) in base).

316

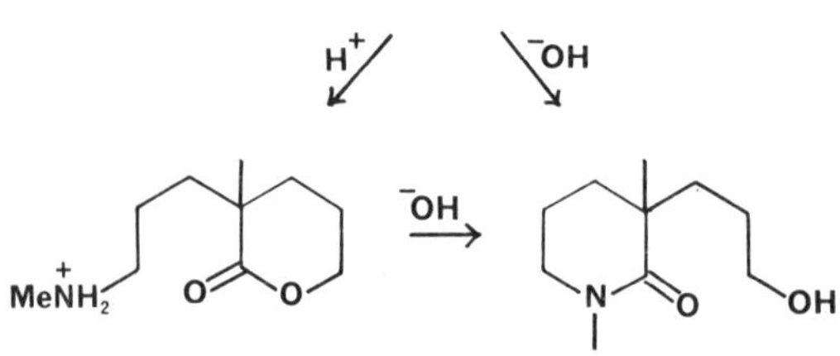

319 318

If (*318*) is in fact the initial product in base, as (*319*) appears to be in acid, both C–O and C–N cleavage must be occurring without stereoelectronic control, presumably by way of twisted conformations of (*317*) and its conjugate acid, respectively, which allow some overlap between the donor and acceptor orbitals. There is strong evidence that this process is much more favourable for donor nitrogen than for oxygen, which would help explain why (*318*) is the preferred product from the base hydrolysis of (*316*).

The best evidence comes from work on rigid systems. Nitrogen is of course intrinsically a better donor than oxygen, because of its lower electronegativity, and covalent α-halogenoamines are too unstable to exist in conformations which could break down with stereoelectronic control. So we cannot hope to compare the rates of cleavage of axial and equatorial compounds, as is possible for the oxygen analogues. But the effects on reactivity of substituting nitrogen for carbon or oxygen in systems like (*320*) and (*321*) are dramatic enough to prove the point.

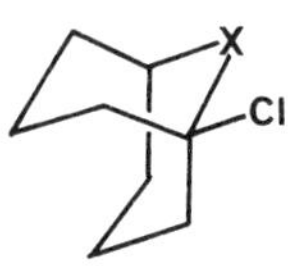

320

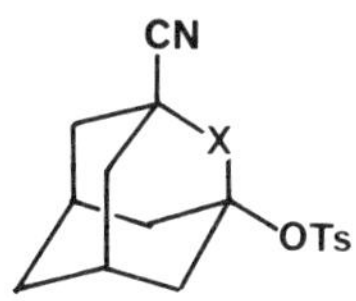

321

We saw above (page 89) that the cleavage of an acetal with the 9-oxo-[3.3.1]-bicyclononane structure as in (*320*, X = O) is subject to a very large stereoelectronic barrier. Compared with a related axial acetal hydrolysis is slower by over ten orders of magnitude, corresponding to a stereoelectronic barrier of 14 kcal mol^{-1}. The rate of solvolysis of the corresponding chloride (*320*, X = O) (96% ethanol at 29 °C) [334] is undoubtedly also much slower than that of a comparable axial chloride, though a direct comparison is not possible. But (*320*, X = O) is also hydrolysed (three times) more slowly than the corresponding tertiary alkyl chloride (*320*, X = CH_2) [334]; evidently the small amount of lone pair donation possible in the transition state for the loss of Cl^- is insufficient even to counteract the inductive effect of an adjacent oxygen atom.

The corresponding aza-compound (*320*, X = NMe), on the other hand, is hydrolysed 8×10^6 times *more rapidly* than the corresponding X = CH_2 compound [334]. Clearly, if electron-demand is strong enough, nitrogen can act as an electron-donor even in this stereoelectronically unfavourable situation. Similar results have been obtained by Meyer and Martin with the oxa- and aza-adamantane systems (*321*), which are significantly more rigid than (*320*) [335]. π-Donation is substantially weaker because of the 3-cyano group as well as the geometry, but in this case also, compared with the X = CH_2 compound the oxygen analogue (*321*, X = O) is less reactive (by a factor of 1340) but the nitrogen derivative (*321*, X = NMe) is (357 times) more reactive [335].

Clearly electron demand is much greater in the cleavage of these halides than is the case for (*317*) and related tetrahedral intermediates T_-; where the leaving group is alkoxide and demand is largely satisfied by donation by O^-. But the transition state for C−O cleavage must have some product (amide) character (though π-overlap cannot develop very far until the C−OR bond is almost completely broken). And the perpendicular conformation (*322*) of an amide, which would be generated by loss of RO^- from a tetrahedral intermediate fixed firmly in conformation (*313*), with no possibility of assistance from the nitrogen lone pair, is less stable than the planar, conjugated form by some 20 kcal mol^{-1} [325a]. In the circumstances it is perhaps suprising that stereoelectronic effects at nitrogen are so small.[15]

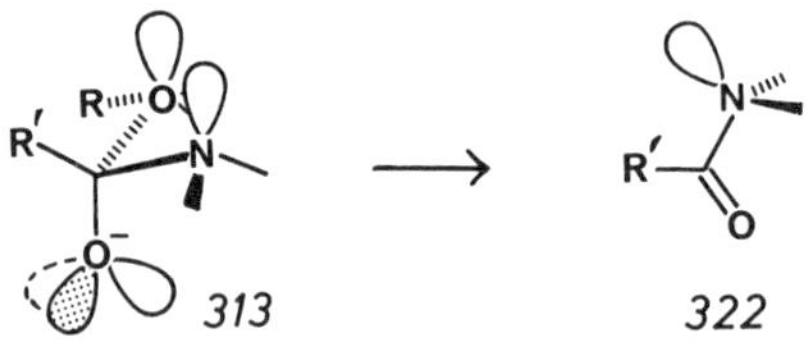

313 322

6. Reactions of Phosphate Esters

Just as calculations on the conformations of dimethyl phosphate revealed close similarities to those for dimethoxymethane (see above, page 56), so calculations on the pentacovalent intermediates involved in displacements at the phosphorus centre of phosphate esters [337–341] (like similar calculations on tetrahedral intermediates [228, 342]) predict that these processes also will be subject to stereoelectronic effects.

The principal effect is once again the weakening of a bond by *antiperiplanar* lone pairs. The energy of the transition state for attack by hydroxide on dimethyl phosphate, for example, is very sensitive to the conformation about the equatorial P−OMe bond [340]. When this is gauche with respect to the bond to the leaving group (*323*) one lone pair on this OMe is able to assist those on the other two equatorial oxygens in breaking the bond to the axial OCH_3.

[15] Significant effects are observed on reactions involving eliminations from C=NR systems [336]. The hydrolysis of (*i*, X = Cl) is 470 times faster than that of the E isomer [336a], and methanolysis generates (*i*, X = OMe),

as expected if the (reverse) addition of nucleophiles to the $C{\equiv}N^+$ triple bond is also stereospecific [336b].

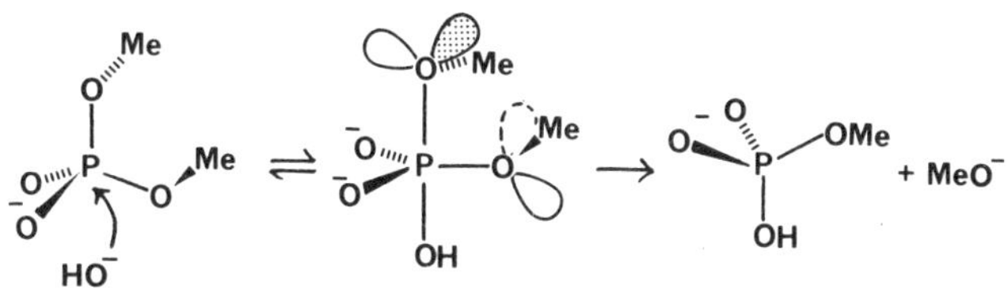

323

The computed transition state energy for the breaking of this bond is substantially lower for this than for other conformations. But it is clear that there is no general requirement for three antiperiplanar lone pairs for the easy breaking of an axial P–O bond, because of the well-known high reactivity of five membered cyclic phosphate esters like ethylene phosphate. Here [338] a conformation corresponding to (*323*) is impossible, but the two

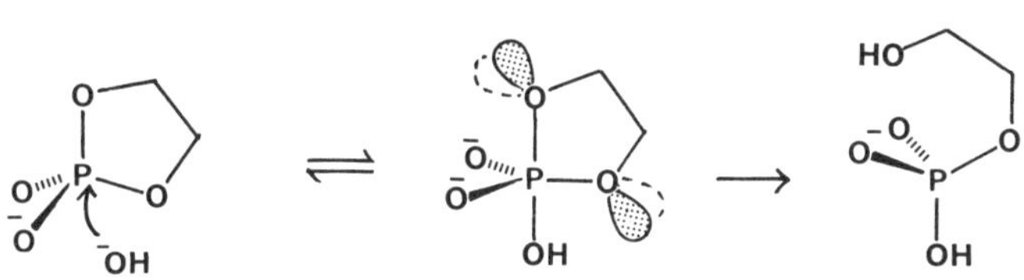

324

anticlinal lone pairs on the equatorial ester oxygen (*324*) are predicted to have an effect comparable to that of one *antiperiplanar* lone pair. (This effect, which is to weaken the apical P–OR bond and *strengthen* the equatorial one, outweighs the reverse effect of the two similar *anticlinal* lone pairs on the axial ester oxygen of (*324*), partly at least because the axial P–O bonds of a pentaalkoxyphosphorane are longer.)

Note that neither pentacovalent intermediate (*323*) or (*324*) has the ideal lone pair geometry for the elimination of OH^- with stereoelectronic control. This is not available at all in the case of a five-membered ring compound (*324*), where the reverse of the addition step must proceed with assistance from lone pairs on the two equatorial P–O groups only. The correct geometry could be arranged in the acyclic system (*323*) by rotation about the equatorial P–OMe bond, if the lifetime of the intermediate is long enough to allow conformational equilibration. If the lifetime is too short (or if the conformation is fixed) either bond making or bond breaking, but not both, can proceed with stereoelectronic control.

Although the results of the calculations on phosphate esters make clear and consistent predictions, there is as yet no good experimental evidence that their reactions are subject to stereoelectronic control. The most relevant work is that of Gorenstein and his group on the bicyclic esters (*325*) [333]; but like the corresponding acetals (see above, page 87) these turn out to be conformationally too mobile to allow a proper test of the theory. In particular the equatorial esters exist largely in [124], and presumably react through [343], the twist-boat conformation (*326*). The transition states for the hydrolysis of the two epimers (*325a* and *e*) (note that addition of hydroxide is rate determining, so there will in any case be little bond *breaking*) are as a result similar in energy, so that reactivity is controlled largely by differences in ground state energy, and the equatorial anomers (*325e*) are consequently the more reactive [343].

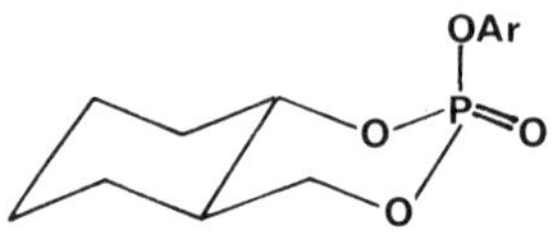

325a

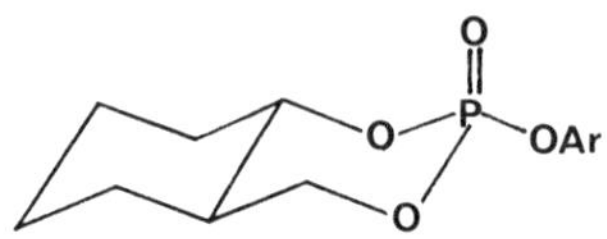

325e

⇌

326

Displacements at pentacovalent phosphorus currently present a complex picture, difficult to rationalise in terms of leaving group ability and conventional conformational properties of molecules alone [344–348]. If such diverse properties as the special reactivity of five-membered cyclic compounds, the kinetically-controlled formation of isomeric pentacovalent and hexacovalent species, and particularly the balance between inversion and retention in displacements at phosphorus, are to be rationalised on a common basis, it may well be in terms of stereoelectronic effects. But so far the necessary synthesis of theory and experiment has not been achieved.

7. Radical Reactions

There is good evidence from recent work that simple radical reactions may be subject to stereoelectronic effects similar to those operating on ionic reactions of acetals and orthoester derivatives. The effects are generally much smaller than those which are observed in ionic reactions of rigid systems; but they are generally more readily observed.

Calculations suggest that the most stable conformation of the tetrahydropyran-2-yl radical (*327*) has the singly occupied molecular orbital (SOMO) axial [349]. This is the optimal geometry for the bonding interaction of the SOMO with the axial lone pair of the ring oxygen, and accounts for the familiar stabilisation of a radical by an adjacent oxygen atom [153b]. (The magnitude of this stabilisation, and its dependence on geometry, is nicely

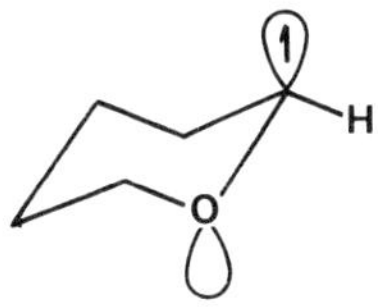

327

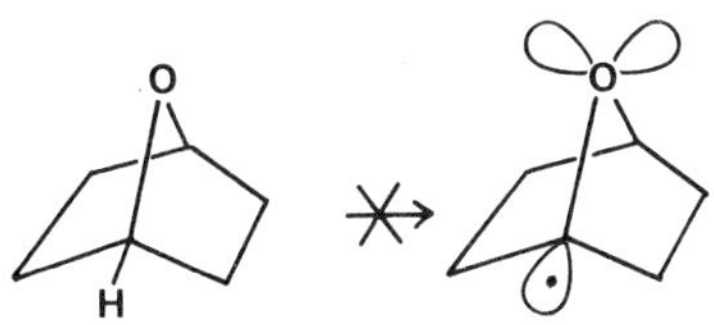

328 329

illustrated by the relative rates of hydrogen abstration from cyclopentane, tetrahydrofuran and the bicyclic ether (*328*): the reactivity of the α-hydrogens (per hydrogen) of tetrahydrofuran is some 3000 times that of those in cyclopentane; but under the same conditions there is no detectable α-hydrogen abstraction (by photochemically generated t-butoxyl) from (*328*); indicating that (*329*), in which overlap between the oxygen lone pairs and the SOMO is prevented, is formed at least an order of magnitude less rapidly [350]. Similar stereoelectronic effects have been observed also for hydrogen abstraction α to the nitrogen atom of tertiary amines [351].

The extra stabilisation afforded by a second oxygen atom appears to be quite small, but equally dependent on geometry. This is shown by the relative rates of hydrogen abstraction by t-BuO˙ from the three tetrahydrofuran acetals (*330–332*) [350]:

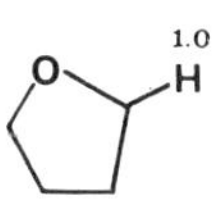

330

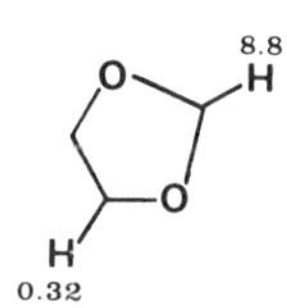

331

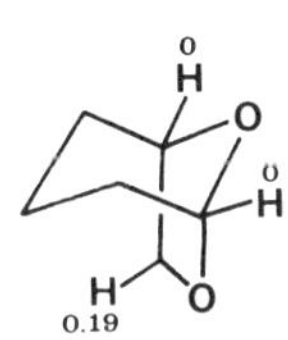

332

which also indicate that an oxygen which cannot stabilise the adjacent radical because of conformational restrictions actually destabilises it.

A more extensive set of data for tetrahydropyran derivatives (Table 24) confirms these conclusions.

Axial hydrogens are removed up to ten times more rapidly than equatorial H from isomeric tetrahydropyran acetals or derived orthoesters; as found earlier by Hayday and McKelvey for the reactions of compounds (*333*) with triplet benzophenone [353]. As in the five-membered ring compounds, the introduction of a second oxygen in a tetrahydropyranyl or open-chain system causes a small increase in reactivity, as long as extra overlap is possible; but not otherwise. But a third oxygen causes no significant increase in reactivity – evidently the stabilisation of a radical by adjacent oxygen rapidly reaches saturation, with a large effect from the first, a small effect from the second, and no significant effect from the introduction of the third

Table 24. Relative Rates of Hydrogen Abstraction by t-BuO[a]

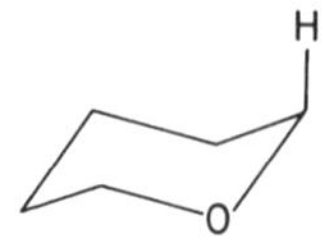	1.0			[350]
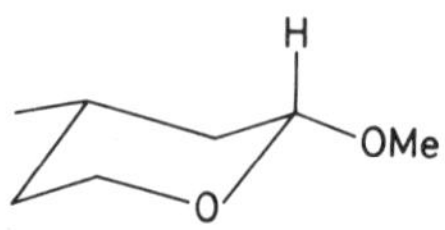 333a	5.9	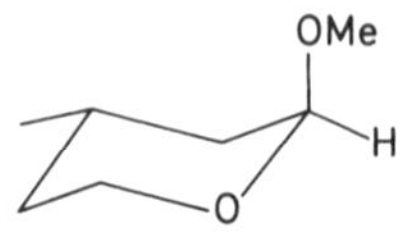 333e	1.5	[350]
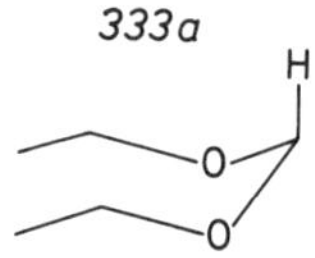	2.6			[350]
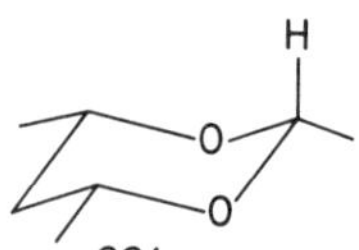 334a	5.6	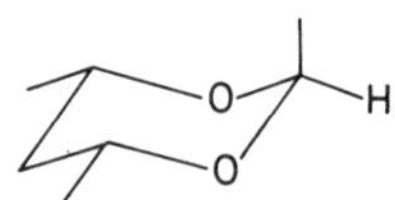 334e	0.5	[352]
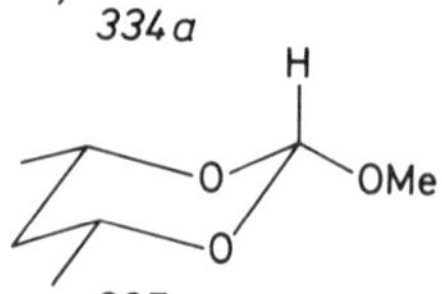 335a	5.0 5.6	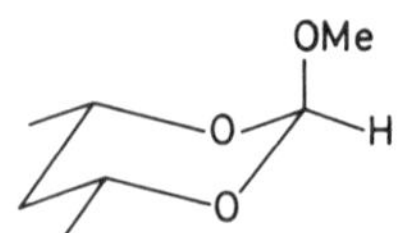 335e	0.6 0.6	[350] [352]
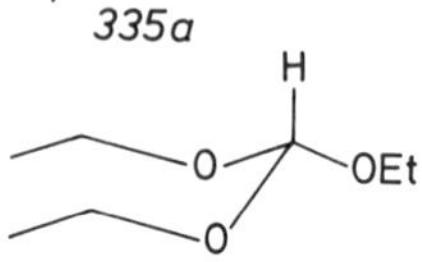	4.8			[350]
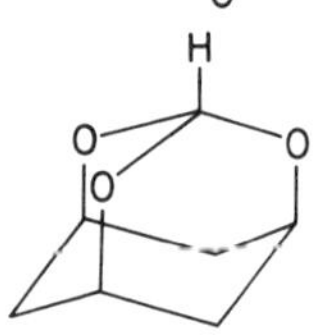 336	not detectable			[350]

[a] Date from Ref. [350] (in CH_2Cl_2 or cyclopropane at $-60°$) and [352] (in CCl_4 at ambient temperatures). Relative reactivities were measured by e.s.r. and g.l.c. methods, respectively, and reactivities have been normalised on the basis of the results for compounds (*335*), which were measured by both methods.

oxygen atom. And once again, if none of the three oxygens can overlap effectively with the SOMO (compound *336*), no reaction is observed.

Other cases where stereoelectronic effects may control reactivity in radical processes are the cleavage of the radicals produced from cyclic acetals, like those from (*333*) or (*337*), which show an apparently general preference for cleavage of the exocyclic bond (which can be *antiperiplanar* to the SOMO (*338*)) to give the lactone rather than the ester product, even where this means generating a methyl rather than a more stable benzyl radical [353, 354] (although reclosure of the benzyl radical (*337*) is a complicating factor): and the preferential loss of axial β-OH groups from inositol (hexahydroxycyclohexane) radicals [355].

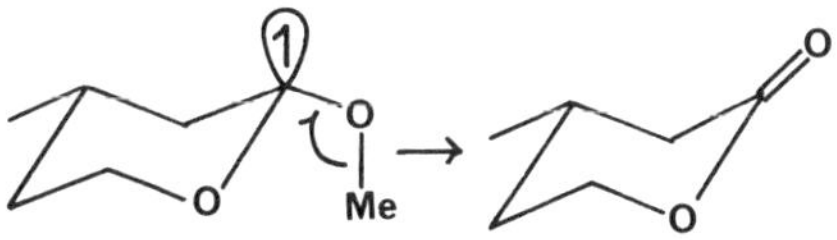

338

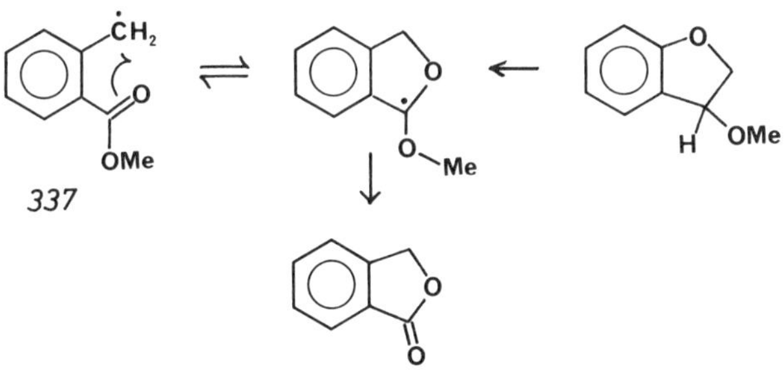

337

The antioxidant activity of α-tocopherol (*339*) (the major E vitamin) also depends on an interesting stereoelectronic factor. The rate determining step in the inhibition of radical reactions by phenols is generally the hydrogen abstraction:

$R^{\cdot} + ArOH \rightarrow ArO^{\cdot} + RH$
$(ArO^{\cdot} + R^{\cdot} \rightarrow$ non-radical products)

This step is over twenty times faster for the vitamin (*339*, R = phytyl; or for R = Me) than for the analogue (*340*). The simple and convincing ex-

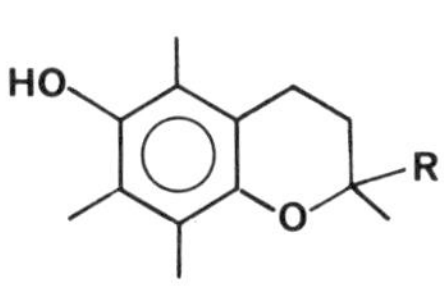

339

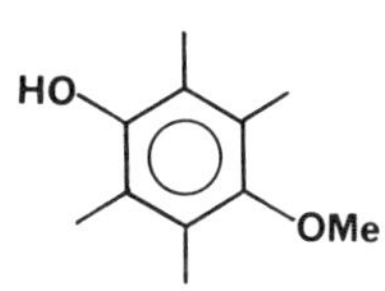

340

planation [356] is that the OMe group of (*340*) is rotated out of plane, so that overlap with the ring is inhibited and the oxygen of the *para*-OR group cannot stabilise the aryloxy radical, as it can in the case of (*341*) from (*339*), where near-planarity is maintained by the ring structure.

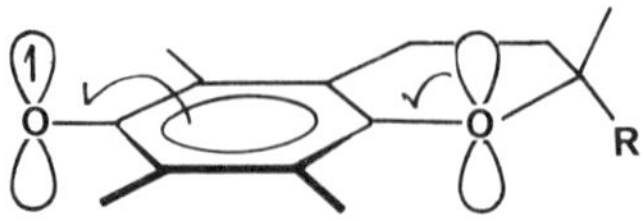

341

Finally, Skell and May [357] have found evidence for acetoxy radicals of two types, distinguished by their different reactivities. Both are generated in reactions initiated by the photolysis of acetal hypobromite. One is the familiar acetoxy radical, which undergoes decarboxylation too rapidly to undergo non-cage reactions, and is likely on several grounds to be the $^2A_1$ σ radical (*342*).

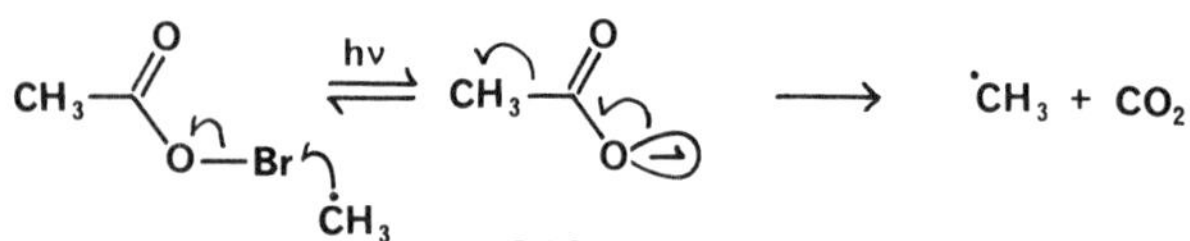

342

The second, formed in the presence of bromine, appears to abstract hydrogen from alkanes with low selectivity in a non-cage reaction. Neither of the two alternative electronic structures (*343, 344*) allows a stereoelectronically favourable pathway for decarboxylation [356].

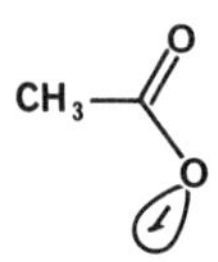

343

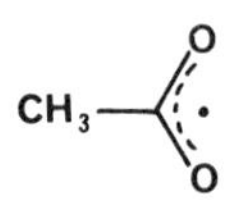

344

8. Enzyme Reactions

The bond making and breaking processes involved in enzyme catalysed reactions are generally so fast that even the very small energy barriers associated with such steps as the diffusion of molecules to or from the reaction site, and conformational changes, become significant, and indeed often rate determining. Under these circumstances it is to be expected that stereoelectronic effects on bond formation and cleavage, and perhaps also on proton transfer steps, will be of particular importance.

This does not, however, mean that they will be readily observable. The disposition of catalytic groups in the active site of a given enzyme will of course take account of stereoelectronic, as of any other factors which have a significant influence on reactivity of its natural substrate. The evidence one can expect from the massive effort involved in demonstrating that the detailed geometry of individual steps of the reaction is consistent with the expectations of stereoelectronic theory cannot therefore prove that stereoelectronic control is involved. And the interpretation of results obtained with modified enzymes or substrates is unlikely to be unambiguous in view of the complexity of even the simplest enzyme-catalysed reactions. Thus there is no compelling evidence of the importance of stereoelectronic effects at oxygen on particular bond making or breaking steps in enzymic reactions.

The best evidence is indirect, and arises in cases where the conformation of the substrate is expected to have a significant effect on reactivity. We have seen, for example, how tetrahydropyranyl acetals with equatorial OR groups are hydrolysed significantly more slowly than the axial anomers if they are held more or less rigidly in the chair conformation (Section C.2(a), above). If the conformation is not fixed reaction occurs by way of an alternative, higher energy conformation, and no stereoelectronic effect is apparent.

This is presumably what happens in glycosyl transfer reactions involving β-glycosides (e.g. *345*) as substrates. In the ground state the bond to the leaving group (OR) is *antiperiplanar* only to ring bonds, so n-donation by the ring oxygen is not possible. Ring inversion would produce a reactive conformation (*347*), but this is scarcely practical for an enzymic process because the first step is substrate binding, which involves recognition of the substrate by the enzyme. Clearly an enzyme which could bind only the all-axial conformation of (*345*) could never be highly efficient because there is no significant amount of substrate in this conformation.

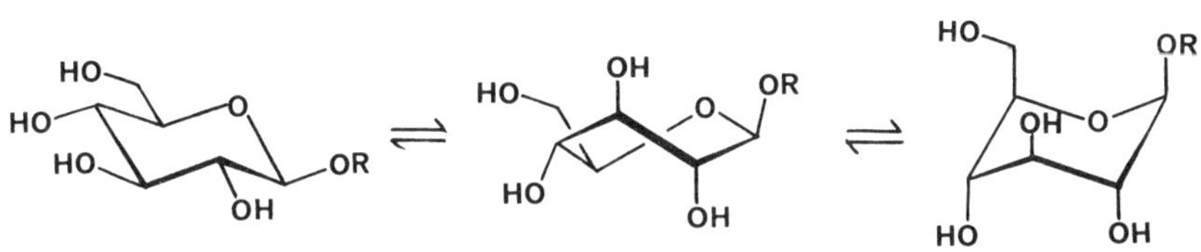

345 *346* *347*

The solution, suggested by Gorenstein and his co-workers [358], and supported by results with model systems, is that reaction goes through a compromise conformation. Enough of the substrate is bound in its ground state conformation for substrate recognition not to be a problem: but binding induces a change in conformation around the reacting glycosidic centre *in the direction of* (*347*), for example to the flexible conformation (*346*), as observed for compounds exhibiting strong anomeric or reverse anomeric effects (see pages 12 and 16, above). This allows n-σ^*_{C-O} overlap, restricted perhaps at first, but more fully developed in the transition state for C−O cleavage.

The classic example, and the original inspiration of much of the work in this area, is the reaction catalysed by the enzyme lysozyme [359]. The natural substrate of this β-glycosidase is an alternating copolymer of N-acetyl-β-D-glucosamine (NAG) and N-acetyl-β-D-muramic acid (NAM) residues, joined by β-1,4-glycosidic bonds, which is a constituent of bacterial cell walls. Comparison of the X-ray structures of the enzyme and its complex with $(NAG)_3$ shows the trisaccharide bound in a cleft in the enzyme, and allows a detailed examination of the hydrogen-bonding and hydrophobic interactions involved in binding.

The trisaccaride fills the cleft for only part of its length, and binding and model-building studies indicate how a substrate as large as a hexasaccharide could be accomodated. Five of the six sugar residues could bind in the stable chair conformation, but the fourth sugar (from the non-reducing end) has to be twisted into a flexible (variously suggested to be a half-chair – as shown in (*348*) – a sofa, or a boat) conformation to allow binding of the fifth and sixth residues. This is highly significant because this is the site (between residues four and five) of the cleavage reaction (*348*→ *349*), and it is difficult to escape the conclusion that the reaction does indeed represent a compromise between the opposing needs of substrate recognition and stabilisation of the developing carbonium ion by the lone pair electrons of the ring oxygen.

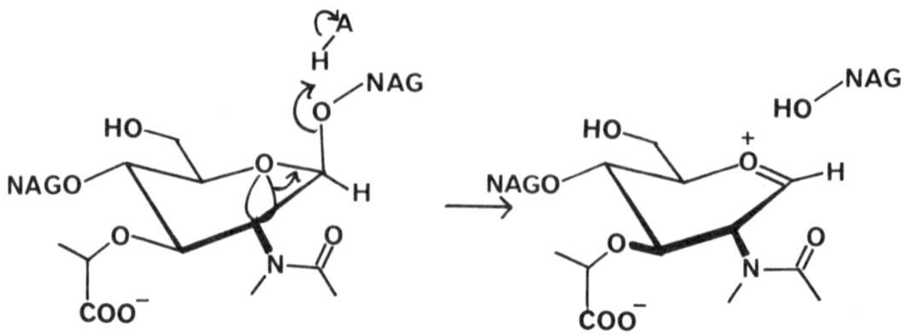

348 *349*

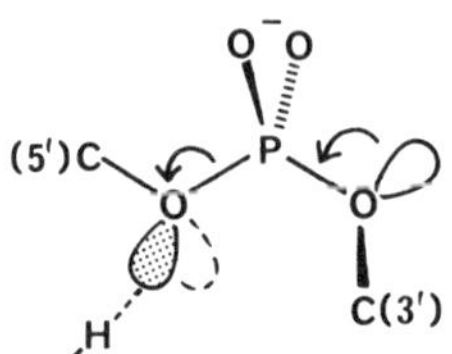

350

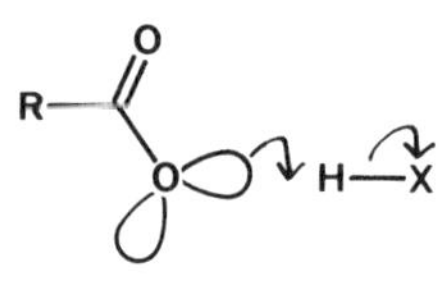

351

Rather similar considerations are involved in the suggestion by Gorenstein et al. that stereoelectronic factors are involved in the specificity of the cleavage of yeast transfer RNA catalysed by ribonuclease A [358]. Of a number of internucleotide 3′,5′-phosphate diester linkages in the double-helical and loop regions of the substrate, where conformations are fairly rigidly fixed, the one cleaved by the enzyme is the only one with the (*sc*, *ap*) conformation (*350*) about the phosphorus centre. (The others are all (+ *sc*, + *sc*).)
In this conformation only the 3′-oxygen can act as a donor, so the bond to be broken is specifically weakened (only slightly of course in the ground state, but the effect can be much larger in the transition state for cleavage

of this bond), and the basicity of the leaving group oxygen (5′) slightly increased, thus facilitating general acid catalysis of its departure. The enzyme thus appears to have 'taken full advantage of the stereoelectronic consequences of the (*sc*, *ap*) conformation' [358].

As we learn more about the details of enzyme mechanisms we may expect this to be the rule. It has been suggested, for example, that the Z lone pair of carboxylate oxygen is effectively more basic than the E-lone pair because the latter is involved in n-σ^* overlap with the other C−O bond [360]. This is unlikely to be a large effect, but Gandour points out that it is the Z lone pair that appears to be generally involved (*351*) in a number of enzyme reactions where carboxylate acts as a general base.

There are perfectly good alternative explanations for all these observations, and as suggested above, a simple and compelling proof of the stereoelectronic explanation is not to be expected. More likely is that the weight of evidence from a large number of systems will eventually prove irresistible.

D. Appendix

(with Dr A.J. Stone)

A comparison of the canonical and hybridised (localised) MO descriptions of the n-σ^*_{C-X} interaction in gauche RO−CH$_2$−X

For representations of the lone pair orbitals, and the conformation referred to, see formulae (*162*) and (*163*) (page 79).

The alternative interaction diagrams as as follows

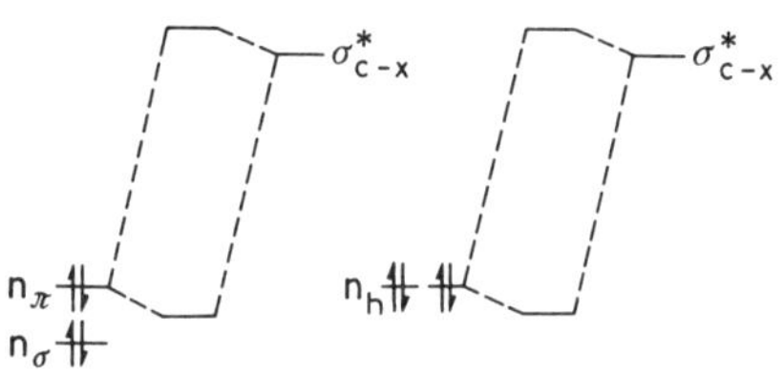

Interaction of the occupied lone pair orbital (n_π or n_h) with the unoccupied σ^*_{C-X} orbital gives a modified orbital $n_i + \lambda\sigma^*_{C-X}$ which is net stabilising, but bonding with respect to C−O and antibonding with respect to C−X.

1. Canonical Lone Pair Description

The secular equation is:

$$\begin{bmatrix} \varepsilon_\sigma - \varepsilon & 0 & V_{\sigma^*} \\ 0 & \varepsilon_\pi - \varepsilon & V_{\pi^*} \\ V_{\sigma^*} & V_{\pi^*} & \varepsilon_* - \varepsilon \end{bmatrix} \begin{bmatrix} c_\sigma \\ c_\pi \\ c_* \end{bmatrix} = 0$$

where ε_σ ε_π and ε_* are the unperturbed energies of the three orbitals. The interactions are described by V_{σ^*} and V_{π^*}.
Since the sum of the eigen values $= \varepsilon_\sigma + \varepsilon_\pi + \varepsilon_{\sigma^*}$
the total decrease in the lone pair energies must equal the increase in the σ^* energy. By second order perturbation theory the latter is

$$\varepsilon - \varepsilon_* = \frac{|V_{\sigma^*}|^2}{\varepsilon_* - \varepsilon_\sigma} + \frac{|V_{\pi^*}|^2}{\varepsilon_* - \varepsilon_\pi} + \ldots \text{(higher order terms)} \quad (1)$$

2. Hybrid Lone Pair Description

Using hybrid lone pairs $h_1 = \sqrt{\frac{1}{2}(\sigma + \pi)}$, $h_2 = \sqrt{\frac{1}{2}(\sigma - \pi)}$, the secular equation becomes:

$$\begin{bmatrix} \varepsilon_h - \varepsilon & V_{12} & V_{1^*} \\ V_{12} & \varepsilon_h - \varepsilon & V_{2^*} \\ V_{1^*} & V_{2^*} & \varepsilon_* - \varepsilon \end{bmatrix} \begin{bmatrix} c_1 \\ c_2 \\ c_* \end{bmatrix} = 0$$

where $\varepsilon_h = \frac{1}{2}(\varepsilon_\sigma + \varepsilon_\pi)$, $V_{12} = \frac{1}{2}(\varepsilon_\sigma - \varepsilon_\pi)$,

$$V_{1^*} = \sqrt{\frac{1}{2}(V_{\sigma^*} + V_{\pi^*})}, \; V_{2^*} = \sqrt{\frac{1}{2}(V_{\sigma^*} - V_{\pi^*})}.$$

Then $$\varepsilon - \varepsilon_* = \frac{|V_{1^*}|^2}{\varepsilon_* - \varepsilon_h} + \frac{|V_{2^*}|^2}{\varepsilon_* - \varepsilon_h} + \ldots \text{(higher order terms)} \quad (2)$$

$$\simeq \frac{|V_{\sigma^*}|^2 + |V_{\pi^*}|^2}{\varepsilon_* - \varepsilon_h} + \ldots \text{(higher order terms)}$$

Thus the only difference to the second order is in the energy terms of the denominators. $\varepsilon_\pi - \varepsilon_\sigma$ is established to be 17 kcal mol^{-1} for tetrahydropyran (*ab initio* calculation) [255], but may be up to twice as large in other oxygen compounds (see above, page 42, and Ref. 157). On the other hand $\varepsilon_* - \varepsilon_\sigma$ is of the order of hundreds of kilocalories: estimates range from about 150 kcal [255] to 380 kcal (for C–F) [193] and even higher [157]. So the denominators in equations (1) and (2) are large numbers, and differ by relatively small amounts (no more than 5%). The quantitative difference between the alternative descriptions is thus quite clearly small enough to be neglected in qualitative, or even semi-qualitative treatments, and, within these limits, the description of the n-σ^* interaction in terms of hybrid lone pairs is equivalent to the formally correct description in terms of canonical molecular orbitals.

E. References

1. Eliel, E.L.: Sterochemistry of Carbon Compounds, especially Chapters 6–8. New York: McGraw Hill 1962
1a. *ibid.*, p. 236
2. Abraham, R.J., Kemp, R.H.: J. Chem. Soc. B *1971*, 1240
3. Jorgensen, W.L., Ibrahim, M.: J. Am. Chem. Soc., *103*, 3976 (1981)
4. Hassel, O., Ottar, B.: Acta Chem. Scand. *1*, 929 (1947)
5. Riddell, F.G.: The Conformational Analysis of Heterocyclic Compounds, p. 66. London and New York: Academic Press 1980
6. Armarego, W.L.F.: Stereochemistry of Heterocyclic Compounds, Part 2, p. 84. New York: Wiley 1977
7. Reeves, R.E.: J. Am. Chem. Soc. *71*, 215 (1949); *72*, 1499 (1950)
8. Angyal, S.J. in The Carbohydrates, Vol. IA, (ed.) Pigman, W. and Horton, D., p. 195. New York: Academic Press 1972
9. Stoddart, J.F.: Stereochemistry of Carbohydrates. New York: Wiley-Interscience 1971. (See also Pople, J.A. in Modern Theoretical Chemistry, Vol. 4, [ed.] Schaefer, H.F., pp. 1–24. New York: Plenum Press 1971)
9a. *ibid.*, p. 172
9b. *ibid.*, p. 72
9c. *ibid.*, p. 101
10. Durette, P.L., Horton, D.: Adv. Carbohydrate Chem. *26*, 49 (1971)
10a. *ibid.*, p. 106
11. Bonner, W.A.: J. Am. Chem. Soc. *81*, 1448 (1959)
12. Lemieux, R.U.: Molecular Rearrangements, Vol. 2, p. 709, (ed.) de Mayo, P.: New York: Interscience 1964
13. Lemieux, R.U., Hayami, J.: Can. J. Chem. *43*, 2162 (1965)
14. Edward, J.T.: Chem. and Ind. *1955*, 1102
15. Lemieux, R.U., Chü, N.J.: Abstr. Papers, Amer. Chem. Soc. *133*, 31N (1958)
16. Martin, J.-C.: Ann. Chim. (Paris) *6*, 205 (1971)
17. Lemieux, R.U. et al.: Can. J. Chem. *47*, 4427 (1969)
18. Szarek, W.A., Horton, D. (eds.) The Anomeric Effect: Origin and Consequences, Am. Chem. Soc. Symposium Series, Vol. 87, Washington 1979
19. Eliel, E.L., Giza, C.A.: J. Org. Chem. *33*, 3754 (1968)
20. Andersen, C.B., Sepp, D.T.: Tetrahedron, *24*, 1707 (1968)
21. Anet, F.A.L.: J. Chem. Soc., *84*, 1053 (1962)
22. Angyal, S.J.: Angew. Chem. Int. Ed. Engl. *8*, 157 (1969)
23. Sepp, D.T., Anderson, C.B.: Tetrahedron *24*, 6873 (1968).
24. Bonner, W.A.: J. Am. Chem. Soc. *81*, 1450 (1959)
25. Thiem, J., Meyer, B., Paulsen, H.P. Chem. Ber. *111*, 3325 (1978)
26. Angyal, S.J.: Aust. J. Chem. *21*, 2737 (1968)
27. Anderson, C.B., Sepp, D.T.: Chem. and Ind. *1964*, 2054
28. Bishop, C.T., Cooper, F.P.: Can. J. Chem. *41*, 2743 (1963)

29. Smirnyagin, V., Cooper, F.P.: Can. J. Chem. *46*, 3085 (1968)
29a. Descotes, G. et al. Compt. Rend. Acad. Sci. Ser. C *267*, 1240 (1968)
30. Pierson, G.O., Runquist, O.A.: J. Org. Chem. *33*, 2572 (1968)
31. de Hoog, A., Havinga, E.: Rec. Trav. Chim. *89*, 972 (1970)
32. Anderson, C.B., Sepp, D.T.: J. Org. Chem. *32*, 607 (1967)
33. Jensen, F.R.: J. Am. Chem. Soc. *91*, 344 (1969)
34. de Hoog, A.J., et al. Tetrahedron *25*, 3365 (1969)
35. Hirsch, J.A.: Top. Stereochem. *1*, 199 (1967)
36. Eliel, E.L., Gilbert, E.C.: J. Am. Chem. Soc. *91*, 5487 (1969)
36a. Fuchs, B., Ellencweig, A. Nouveau J. de Chim. *3*, 145 (1979)
37. Horton, D., Turner, W.N.: J. Org. Chem. *30*, 3387 (1965)
38. Holland, C.V., Horton, D., Jewell, J.S.; J. Org. Chem. *32*, 1818 (1967)
39. Hall, L.D., Manville, J.F.: Can. J. Chem. *47*, 1 (1969)
40. Hall, L.D., Manville, J.F.: Can. J. Chem. *47*, 19 (1969)
41. Luger, P., Durette, P.L., Horton, D.: Chem. Ber. *107*, 2616 (1974)
42. Paulsen, H., Luger, P., Heiker, F.R.: Ref. 18, pp. 63–79
43. Durette, P.L., Horton, D.: J. Org. Chem. *36*, 2658 (1971)
44. Durette, P.L., Horton, D.: Carbohydrate Research, *18*, 403 (1971)
45. Luger, P., et al. Carbohydrate Research *68*, 207 (1979)
46. Luger, P., Kothe, G., Paulsen, H.: Chem. Ber. *109*, 1850 (1976)
47. Luger, P., Kothe, G., Paulsen, H.: Angew. Chem. Int. Ed. Engl. *16*, 52 (1977)
48. Vangehr, K., Luger, P., Paulsen, H.: Chem. Ber. *113*, 2609 (1980)
49. Hall, L.D., Manville, J.F.: Can. J. Chem. *47*, 1 (1969)
50. Bhacca, N.S., Horton, D., Paulsen, H.: J. Org. Chem. *33*, 2484 (1968)
51. Lemieux, R.U., Pavia, A.A.: Can. J. Chem. *47*, 4441 (1969)
52. Isbrandt, L.R., Oertel, R.P.: J. Am. Chem. Soc. *102*, 3144 (1980)
53. Bailey, W.F., Eliel, E.L.: J. Am. Chem. Soc., *96*, 1798 (1974)
54. Paulsen, H., Györgydeak, Z., Friedmann, M.: Chem. Ber. *107*, 1590 (1974)
55. Lemieux, R.U., Morgan, A.R.: Can. J. Chem. *43*, 2205 (1965)
56. James, M.N.G.: Proc. Can. Fed. Biol. Soc. *13*, 71 (1969)
57. Depmeier, W., von Voithenburg, H., Jochims, J.C.: Chem. Ber. *111*, 2010 (1978)
58. Lemieux, R.U.: Pure Appl. Chem. *25*, 527 (1971)
59. Onodera, K., Hirano, S., Masuda, F.: Carbohydrate Research *7*, 27 (1968)
60. Finch, P., Nagpurkar, A.G.: Carbohydrate Research *49*, 275 (1976)
61. Doland, L. et al.: Eur. J. Biochem. *73*, 93 (1977)
62. de Bruyn, A. et al.: Bull. Chem. Soc. Belg. *87*, 783 (1979)
63. Coxon, B.: Tetrahedron *22*, 2281 (1966)
64. Anderson, C.B., Sepp, D.T.: J. Org. Chem. *33*, 3272 (1968)
65. Lemieux, R.U., Koto, S., Voisin, D.: Ref. 18, p. 17
66. Lemieux, R.U.: Ann. N.Y. Acad. Sci. *222*, 915 (1973)
67. Lemieux, R.U., Koto, S.: Tetrahedron *30*, 1933 (1974)
68. Lemieux, R.U., Nagabhushan, T.L., Paul, B.: Can. J. Chem. *50*, 773 (1972)
69. Perez, S., Marchessault, R.H.: Carbohydrate Research *65*, 114 (1978)
70. Lemieux, R.U., Martin, J.-C.: Carbohydrate Research *13*, 139 (1970)
71. Pavia, A.A., Ung-Chhun, S.N., Lacombe, J.-M.: Nouveau J. de Chim. *5*, 101 (1981)
72. Bundle, D.R., Josephson, S.: Can. J. Chem. *57*, 662 (1979)
73. Painter, T.J.: J. Chem. Soc. Perkin 2 *1976*, 215
74. Rees, D.A., Scott, W.E.: J. Chem. Soc. B *1971*, 469
75. Tvaroška, I., Kožár, T.: J. Am. Chem. Soc. *102*, 6929 (1980)
76. Deslongchamps, P. et al.: Can. J. Chem. *59*, 1105 (1981)
77. Pothier, N. et al.: Can. J. Chem. *59*, 1132 (1981)
78. Evans, D.D. et al.: Tetrahedron Letters *1978*, 727

79. Hughes, D.L.: Tetrahedron Letters *1978*, 3959
80. Aoki, K.: Nippon Kagaku Zasshi *74*, 110 (1953). Astrup, E.E.: Acta Chem. Scand. *25*, 1494 (1971)
81. Christ, H.A.: Helv. Chim. Acta *33*, 572 (1960)
82. Brown, G.M., Levy, H.A.: Oak Ridge Nat. Lab. Chem. Ann. Rept. 1962
83. Uchida, T., Kurita, Y., Kubo, M.: J. Polymer Sci. *19*, 365 (1966)
84. Hirota, H.: J. Mol. Spectrosc. *26*, 335 (1968). Bolton, K., Owen, N.L. and Sheridan, J.: Nature (London) *217*, 164 (1968)
85. Eliel, E.: Angew. Chem. Int. Ed. Engl. *11*, 739 (1972)
86. Zefirov, N.S.: Tetrahedron *33*, 3193 (1977)
87. Eliel, E.L. et al.: Conformational Analysis, p. 460, New York: Wiley 1965
88. Bernasconi, C.F. et al.: Nouveau J. de Chim. *2*, 79 (1978)
89. Lessard, J. et al.: Can. J. Chem. *55*, 1015 (1977)
90. Booth, H., Lemieux, R.U.: Can. J. Chem. *49*, 777 (1971).
91. Katritzky, A.R., Brito-Palma, F.M.S., Baker, V.J.: J. Chem. Soc. Perkin 2 *1980*, 1739
92. Hutchins, R.O., Kopp, L.D., Eliel, E.L.: J. Am. Chem. Soc. *90*, 7174 (1968)
93. Katritzky, A.R. et al.: J. Chem. Soc. Perkin 2 *1980*, 1745
94. Baker, V.J. et al.: J. Chem. Soc. Perkin 2 *1978*, 377. Duke, R.P. et al.: *ibid. 1973*, 2109. Bushweller, C.H., Lourandos, M.Z., Brunelle, J.A.: J. Am. Chem. Soc. *96*, 1591 (1974)
95. Jones, R.A.Y., Katritzky, A.R., Snarey, M.: J. Chem. Soc. B *1970*, 135
95a. Riediker, M., Graf, W.: Helv. Chim. Acta *62*, 2053 (1979)
96. Romers, C. et al.: Top. Stereochem. *4*, 39 (1967)
97. Kalff, H.T., Romers, C.: Acta Cryst. *18*, 164 (1965). Rec. Trav. Chim. *85*, 198 (1966)
98. de Wolf, N., Romers, C., Altona, C.: Acta Cryst. *22*, 715 (1967), de Hoog, A.J., Havinga, E.: Rec. Trav. Chim. *89*, 972 (1970)
99. de Hoog, A.J.: Thesis (Leiden) 1971
100. Zefirov, N.S.: J. Org. Chem. USSR *5*, 1131 (1969)
101. Zefirov, N.S.: Zh. Org. Khim. *1*, 594 (1971)
102. Oki, M., Sugawara, T., Iwamura, H.: Chem. Letters *1973*, 997. idem., Bull. Chem. Soc. Japan *47*, 2457 (1974)
103. Oki, M., Sugawara, T., Iwamura, H.: Bull. Chem. Soc. Japan *47*, 1496 (1974)
104. Arai, K., Iwamura, H., Oki, M.: Bull. Chem. Soc. Japan *48*, 3319 (1975)
105. Deslongchamps, P. et al.: Can. J. Chem. *59*, 1122 (1981)
106. Featherman, S.C., Quin, L.D.: J. Am. Chem. Soc. *97*, 4349 (1975)
107. McPhail, A.T. et al.: J. Am. Chem. Soc. *94*, 2126 (1974)
108. MacDonell, G.L. et al.: J. Am. Chem. Soc. *100*, 4535 (1978)
109. White, D.W. et al.: J. Am. Chem. Soc. *92*, 7125 (1970)
110. Bentrude, W.G., Tan, H.-W., Yee, K.C.: J. Am. Chem. Soc. *97*, 573 (1975)
111. Mosbo, J.A., Verkade, J.G.: J. Am. Chem. Soc. *95*, 4659 (1975)
112. Stec, W. et al.: J. Chem. Soc. Perkin 1 *1975*, 1828
113. Hutchins, R.O., Maryanoff, B.E.: J. Am. Chem. Soc. *94*, 3266 (1972)
114. Radom, L., Hehre, W.J., Pople, J.A.: J. Am. Chem. Soc. *94*, 2371 (1972)
115. Hall, L.D., Malcolm, R.B.: Can. J. Chem. *50*, 2092 (1972)
116. Geise, H.J.: Rec. Trav. Chim. *86*, 362 (1967)
117. Bodkin, C.L., Simpson, P.: J. Chem. Soc. Chem. Commun. *1969*, 829
118. Mosbo, J.A., Verkade, J.G.: J. Org. Chem. *52*, 1549 (1977)
119. Bentrude, W.G., Hargis, J.H., J. Am. Chem. Soc. *92*, 7136 (1970)
120. Majoral, J.-P., Bergounhou, G., Navech, J.: Bull. Soc. Chim. France *1973*, 3146
121. Warrent, R. et al.: J. Org. Chem. *43*, 4266 (1978)

122. Clardy, J.C., Mosbo, J.A., Verkade, J.G.: J. Chem. Soc. Chem. Commun. *1972*, 1163. Garcia-Blanco, S., Perales, A.: Acta Cryst. *B28*, 2647 (1972)
123. Bajwa, G.S. et al.: J. Am. Chem. Soc. *101*, 1602 (1979). Chandrasekhar, S., Bentrude, W.G.: Tetrahedron Letters *1980*, 4671
124. Gorenstein, D.G., Rowell, R.: J. Am. Chem. Soc. *101*, 4925 (1979). Gorenstein, D.G., Rowell, R., Findlay, J.B.: *ibid. 102*, 5077 (1980)
125. Bentrude, W.G., Tan, H.-W.: J. Am. Chem. Soc. *95*, 4666 (1973)
126. Maryanoff, B.E., Hutchins, R.O.: J. Org. Chem. *42*, 1022 (1977)
127. Hutchins, R.O. et al.: J. Am. Chem. Soc. *101*, 1600 (1979). Maryanoff, B.E., Huthins, R.O., McPhail, A.T.: J. Am. Chem. Soc. *103*, 4432 (1981)
128. Eliel, E.L. et al.: J. Am. Chem. Soc. *96*, 3021 (1974)
129. Bergesen, K., Carden, B.M., Cook, M.J.: J. Chem. Soc. Perkin 2, *1978*, 1001
130. Gelan, J., Swaelens, G., Anteunis, M.: Bull. Chem. Soc. Belges *79*, 321 (1970)
131. Buchanan, G.W. et al.: Can. J. Chem. *55*, 44 (1977)
132. Maroni, P. et al.: Bull. Soc. Chim. France *1975*, 1253
133. van Woerden, H.F., Havinga, E.: Rec. Trav. Chim. *86*, 341, 353 (1967)
134. Wood, G.W., McIntosh, J.M., Miskow, M.A.: Can. J. Chem. *49*, 1202 (1971): *50*, 521 (1972)
135. Buchanan, G.W., Stothers, J.B., Wood, G.W.: *ibid. 51*, 3746 (1973)
136. Altona, C., Geise, H.J., Romers, C.: Rec. Trav. Chim. *85*, 1197 (1966). van Oyen, J.W.L. et al.: Acta Cryst. *B24*, 1471 (1968)
137a. Exner, O., Fidlerová, Z., Jehlička, V.: Coll. Czech. Chem. Commun. *33*, 2019 (1968).
137b. Exner, O., Dembach, P., Vivarelli, P.: J. Chem. Soc. B *1970*, 278
138. Remizov, A.B., Butenko, G.G.: Zh. Prikl. Spektrosk. *20*, 545 (1974)
139. Klimovitskii, E.N. et al.: Izv. Akad. Nauk SSSR Ser. Khim. *1977*, 98
140. Winnewisser, G., Winnewisser, M., Gordy, W.: J. Chim. Phys. *49*, 3465 (1968)
141. Wolfe, S.: Accts. Chem. Res. *5*, 102 (1972)
142. Nelsen, S.F.: Accts. Chem. Res. *11*, 14 (1978)
143. Wolfe, S. et al.: J. Chem. Soc. B *1971*, 136
144. Phillips, L., Wray, V.: J. Chem. Soc. Chem. Commun. *1973*, 90
145. Planje, M.C., Toneman, L.H., Dallinga, G.: Rec. Trav. Chim. *84*, 232 (1965). Aroney, M.J., Le Fèvre, R.J.W., Saxby, J.D.: J. Chem. Soc. B *1966*, 414
146. Kaloustian, M.K. et al.: J. Am. Chem. Soc. *98*, 956 (1976)
147. Kotite, M.J., Harris, M., Kaloustian, M.K.: J. Chem. Soc. Chem. Commun. *1977*, 911
148. Eliel, E.L., Juaristi, E.: J. Am. Chem. Soc. *100*, 6114 (1978)
149. Eliel, E.L., Juaristi, E.: Ref. 18, p. 95
150. Zefirov, N.S. et al.: Tetrahedron *32*, 1211 (1976)
150a. Zefirov, N.S.: Zh. Org. Khim. *6*, 1761 (1970). Zefirov, N.S. et al.: Tetrahedron *27*, 3111 (1971)
151. Gerlt, J.A. et al.: J. Am. Chem. Soc. *102*, 1655, 1660, 1665 (1980)
151a. Altona, C., Havinga, E.: Tetrahedron *22*, 2275 (1966)
151b. Fuchs, B. et al.: Nouveau J. de Chim. *3*, 145 (1979). Fuchs, B., Auerbach, Y., Sprecher, M.: Tetrahedron *30*, 437 (1974)
152. Altona, C.: Thesis, Leiden, 1964. Altona, C., Romers, C.: Acta Cryst. *16*, 225 (1963)
153. Fleming, I.: Frontier Orbitals and Organic Chemical Reactions, p. 7, New York and London: Wiley 1976
154. Kirby, A.J., Warren, S.G.: The Organic Chemistry of Phosphorus, p. 3, Amsterdam: Elsevier 1967
154a. Kost, D.L. et al.: Can. J. Chem. *57*, 729 (1979)

155. Robb, M.A., Heimes, W.J., Csizmadia, I.G.: J. Am. Chem. Soc. *95*, 42 (1973)
156. Lien, M.H. et al.: Progr. Theor. Org. Chem. vol. 2, (ed.) Csizmadia, I.G. et al.: p. 162, Amsterdam: Elsevier 1977. Wolfe, S. et al.: J. Am. Chem. Soc. *95*, 4863 (1973)
157. Jorgenson, W.L., Salem, L.: The Organic Chemist's Book of Orbitals, p. 42, New York: Academic Press 1973
158. David, S. et al.: J. Am. Chem. Soc. *95*, 5044 (1973)
159. Dunning, T.H., Pitzer, R.M., Aung, S.: J. Chem. Phys. *57*, 5044 (1972)
160. Allinger, N.L., Chung, D.Y.: J. Am. Chem. Soc. *98*, 6798 (1976)
161. Atoji, M., Lipscomb, W.N.: Acta Cryst. *7*, 173 (1954)
162. Dulmage, W.J., Lipscomb, W.N.: Acta Cryst. *4*, 330 (1951)
163. Pauling, L.: The Nature of the Chemical Bond, p. 464, Ithaca, Cornell University Press 1960 163a. *ibid.* p. 498
164. Whalley, E. in The Hydrogen Bond, vol. III (ed.) Schuster, P., Zundel, G., Sandorfy, C., p. 1425, Amsterdam: North Holland 1976
165. Donohue, J. in Structural Chemistry and Molecular Biology (ed.) A. Rich, A., Davidson, N.: p. 443, San Francisco: Freeman 1968. Acta Cryst. *B24*, 1558 (1968)
166. Olovsson, I., Jönsson, P.-G. in Ref. 164, vol. II, p. 418
167. Kroon, J. et al.: J. Mol. Struct. *24*, 109 (1975)
168. Ceccarelli, C., Jeffrey, G.A., Taylor, R.: J. Mol. Struct. *70*, 255 (1981)
169. Kollman, P.A.: J. Am. Chem. Soc. *94*, 1837 (1972)
170. Dunitz, J.D.: X-Ray Analysis and the Structure of Organic Molecules, pp. 391 bis 417, Ithaca: Cornell University Press 1979
171. Coppens, P.: Angew. Chem. Int. Ed. Engl. *16*, 32 (1977)
172. Coppens, P., Lehmann, M.S.: Acta Cryst. *B32*, 1777 (1976)
173. Berkovich-Yellin, Z., Leiserowitz, L.: J. Am. Chem. Soc. *97*, 5627 (1975)
174. Irngartinger, H., Leiserowitz, L., Schmidt, G.M.J.: J. Chem. Soc. B *1970*, 497
175. Becker, P., Coppens, P., Ross, F.K.: J. Am. Chem. Soc. *95*, 7604 (1973)
176. Verschoor, G.C., Keulen, E.: Acta Cryst. *B37*, 134 (1971). Coppens, P. and Vos, A.: *ibid.* 146
177. van der Wal, H.R., Vos, A.: Acta Cryst. *B35*, 1793, 1804 (1979)
178. Hirshfeld, F.L., Hope, H.: Acta Cryst. *B36*, 406 (1980)
179. Eisenstein, M.: Acta Cryst. *B35*, 2614 (1979)
180. Stevens, E.D., Coppens, P.: Acta Cryst. *B36*, 1864 (1980)
181. Stevens, E.D., Rys, J., Coppens, P.: J. Am. Chem. Soc., *100*, 2324 (1978)
182. Thomas, J.O.: Acta Cryst. *B33*, 2867 (1977)
183. Tellgren, R., Thomas, J.O., Olovsson, I.: Acta Cryst. *B33*, 3500 (1977)
184. Bye, E., Dunitz, J.D.: Acta Cryst. *B34*, 3245 (1978)
185. Amstutz, R., Dunitz, J.D., Seebach, D.: Angew, Chem. Int. Ed. Engl. *20*, 465 (1981)
186. Maverick, E. et al.: Acta Cryst. *B36*, 615 (1980)
187. Hanson, J.C., Sieker, L.C., Jensen, L.H.: Acta Cryst. *B29*, 797 (1973)
188. Stevens, E.D., Coppens, P.: Acta Cryst. *B36*, 1864 (1980)
189. Savariault, J.-M., Lehmann, M.S.: J. Am. Chem. Soc. *102*, 1298 (1980)
190. Jeffrey, G.A. et al.: J. Am. Chem. Soc. *100*, 373 (1978)
191. Gorenstein, D.G., Kar, D.: J. Am. Chem. Soc. *99*, 672 (1977)
192. Tvaroska, I., Bleha, T.: J. Mol. Struct. *24*, 249 (1975)
193. Wolfe, S., Whangbo, M.-H., Mitchell, D.J.: Carbohydrate Research *69*, 1 (1979)
194. Jeffrey, G.A., Pople, J.A., Radom, L.: Carbohydrate Research *25*, 117 (1972)
195. Jeffrey, G.A., Pople, J.A., Radom, L.: Carbohydrate Research *38*, 81 (1974)
196. Jeffrey, G.A., Yates, J.H.: Carbohydrate Research (in press)
197. Jeffrey, G.A., Yates, J.H.: Carbohydrate Research *79*, 155 (1980)
198. Jeffrey, G.A., Yates, J.H.: J. Am. Chem. Soc. *101*, 820 (1979)

198a. Anet, F.A.L.: J. Am. Chem. Soc. *99*, 6752 (1977)
199. Brockway, L.O.: J. Phys. Chem. *41*, 185 (1937)
200. Lide, D.R.: J. Am. Chem. Soc. *74*, 3548 (1952)
201. Astrup, E.: Acta Chem. Scand. *27*, 327 (1973)
202. Lees, R.M., Baker, J.G.: J. Chem. Phys. *48*, 5299 (1968)
203. Blukis, U., Kasai, P.H., Myers, R.J.: J. Chem. Phys. *38*, 2753 (1963)
204. Hayashi, M., Kato, H.: Bull. Chem. Soc. Japan *53*, 2701 (1980)
205. Williams, J.O. et al.: J. Mol. Struct. (Theochem.) *76*, 11 (1981)
206. Bürgi, H.B., Lehn, J.-M., Wipff, G.: Helv. Chim. Acta *57*, 493 (1974)
207. Gorenstein, D.G. et al.: J. Am. Chem. Soc. *99*, 3473 (1977)
208. Gorenstein, D.G., Kar, D.: J. Am. Chem. Soc. *99*, 672 (1977)
209. Tvaroska, I., Bleha, T.: Can. J. Chem. *57*, 424 (1979)
210. Jeffrey, G.A., in Ref. 18, p. 50
211. Herpin, P. et al.: Acta Cryst, *B32*, 209 (1976)
212. Kothe, G., Luger, P., Paulsen, H.: Carbohydrate Research *37*, 283 (1974)
213. Brown, K.L. et al.: Acta Cryst. (in press)
214. Kirby, A.J.: Adv. Phys. Org. Chem. *17*, 183 (1980)
215. Jones, P.G., Kirby, A.J.: J. Chem. Soc. Chem. Commun. *1979*, 288
216. Jones, P.G., Kirby, A.J.: unpublished work
217. Koeners, H.J. et al.: Rec. Trav. Chim. *99*, 355 (1980)
218. Jones, P.G. et al.: Acta Cryst. *B34*, 2947 (1978)
219. Jones, P.G. et al.: Acta Cryst. *B35*, 242 (1979)
220. Jones, P.G. et al.: Acta Cryst. *B35*, 755 (1979)
221. Gerlt, J.A., Youngblood, A.V.: J. Am. Chem. Soc. *102*, 7433 (1980)
222. Lo, A., Shefter, E., Cochran, T.G.: J. Pharm. Sci. *64*, 1707 (1975)
223. de Leeuw, H.P.M., Haasnoot, C.A.G., Altona, C.: Isr. J. Chem. *20*, 108 (1980)
224. Egert, E. et al.: J. Am. Chem. Soc. *102*, 3707 (1980)
225. Epiotis, N. et al.: Structural Theory of Organic Chemistry, pp. 49–52, Berlin–Heidelberg–New York: Springer 1977
226. Eisenstein, O. et al.: Tetrahedron *30*, 1717 (1974)
227. Cottier, L. et al.: Acta Cryst. *B37*, 1155 (1981)
228. Lehn, J.-M., Wipff, G.: J. Am. Chem. Soc. *102*, 1347 (1980)
229. Simonetta, M., Carra, S.: in The Chemistry of Carboxylic Acids and Esters, (ed.) Patai, S., p. 1, London and New York: Wiley 1969
230. Curl, R.F.: J. Chem. Phys. *30*, 1529 (1959)
231. True, N.S., Bohn, R.K.: J. Am. Chem. Soc. *98*, 1188 (1976)
232. Hopkinson, A.C., Yates, K., Csizmadia, I.G.: J. Chem. Phys. *52*, 1784 (1970)
233. Radom, L. et al.: Aust. J. Chem. *25*, 1601 (1972)
234. Deslongchamps, P.: Heterocycles *7*, 1271 (1977): Tetrahedron *31*, 2461 (1975): Pure Appl. Chem. *43*, 351 (1975)
235. Larson, J.R., Epiotis, N.D., Bernardi, F.: J. Am. Chem. Soc. *100*, 5713 (1978)
236. Fodor, G., Phillips, B.A.: in The Chemistry of Amidines and Imidates, (ed.) Patai, S., p. 132. London/New York: Wiley 1975
237. Moriarty, R.M. et al.: J. Am. Chem. Soc. *92*, 6360 (1970)
238. Hegarty, A.F. et al.: J. Chem. Soc. Perkin 2 *1980*, 867
239. McCarthy, D.G., Hegarty, A.F.: J. Chem. Soc. Perkin 2, *1977*, 1080, 1085
240. Brady, K., Hegarty, A.F.: J. Chem. Soc. Perkin 2 *1980*, 121. Meese, C.O., Walter, W., Berger, M.: J. Am. Chem. Soc. *96*, 2259 (1974)
241. Johnson, J.E. et al.: J. Org. Chem. *46*, 546 (1980)
241a. Lee, I., Park, D.W.: Taehan Hwahakhoe Chi *23*, 358 (1979): Chem. Abs. *93*, 94602 (1980)
242. Deslongchamps, P., Lebreux, C., Taillefer, R.J.: Can. J. Chem. *51*, 1665 (1973)

243. Mersh, J.D., Sanders, J.K.M.: Tetrahedron Letters *1981*, 4029
244. Jost, R. et al.: J. Am. Chem. Soc. *102*, 7663 (1980)
245. Jones, G.Y.L., Owen, N.L.: J. Mol. Struct. *18*, 1 (1973)
246. Mizushima, Z., Kubo, H.: Bull. Chem. Soc. Japan *13*, 174 (1938)
247. Bock, E., Iwacha, D.: Can. J. Chem. *45*, 3177 (1967)
248. Queen, A., Nour, T.A., Bock, E.: Can. J. Chem. *47*, 343 (1969)
249. Bock, E. et al.: Can. J. Chem. *46*, 1645 (1968)
250. Jackson, R.H.: J. Chem. Soc. *1962*, 4585
251. Epiotis, N.D. et al.: J. Am. Chem. Soc. *99*, 8379 (1977)
252. Brunck, T.K., Weinhold, F.: J. Am. Chem. Soc. *101*, 1700 (1979)
253. Gavezzotti, A., Bartell, L.S.: J. Am. Chem. Soc. *101*, 5142 (1979)
254. Traylor, T.G. et al.: J. Am. Chem. Soc. *93*, 5715 (1971)
255. David, S. et al.: J. Am. Chem. Soc. *94*, 6221 (1972)
256. David, S., ref. 18, p. 1. Lehn, J.-M., Wipff, G.: J. Am. Chem. Soc. *98*, 7498 (1976)
257. McKean, D.C. et al.: Chem. Soc. Rev. *1978*, 399
258. Steigerwald, M.L., Goddard, W.A., Evans, D.A.: J. Am. Chem. Soc. *101*, 1994 (1979)
259. Blom, C.E.: Thesis: Leiden, 1976
260. Petrzilka, M., Felix, D., Eschenmoser, A.: Helv. Chim. Acta *56*, 2950 (1973)
261. David, S., Ref. 18, p.1
262. Erhardt, J.M., Grover, E.R., Wuest, J.D.: J. Am. Chem. Soc. *102*, 6365 (1980)
262a. Weisman, G.R., Johnson, V.N., Coolidge, M.B.: Tetrahedron Letters *1981*, 4365. Vierhapper, F.W., Eliel, E.L., Zúñiga, G.: J. Org. Chem. *45*, 4844 (1980)
263. Bach, R.D., Badger, K.C., Lang, T.J.: J. Am. Chem. Soc. *101*, 2845 (1979)
264. Bürgi, H.B., Lehn, J.-M., Wipff, G.: J. Am. Chem. Soc. *96*, 1956 (1973)
265. Farcasiu, D., Horsley, J.A.: J. Am. Chem. Soc. *102*, 4906 (1980)
266a. Ha, T.-K.: Chem. Phys. Letters *30*, 379 (1975)
266b. Ros, P.: J. Chem. Phys. *49*, 4902 (1968)
267. Ferrier, R.J., Collins, P.M.: Monosaccharide Chemistry, p. 58, London: Penguin 1972
268. Capon, B.: Chem. Rev. *69*, 407 (1969)
269. Capon, B., Thacker, D.: J. Chem. Soc. B *1967*, 1010. Banks, B.E.C. et al.;
270. J. Chem. Soc. 1961, 3240
Fife, T.H., Brod, L.H.: J. Am. Chem. Soc. *92*, 1681 (1970). Fife, T.H., Jao, L.K.: J. Am. Chem. Soc. *90*, 4081 (1968)
271. Craze, G.-A., Kirby, A.J.: J. Chem. Soc. Perkin 2 *1978*, 354
272. Chandrasekhar, S., Kirby, A.J.: J. Chem. Soc. Chem. Commun. *1978*, 171
273. Saunders, W.H., Cockerill, A.F.: Mechanisms of Elimination Reactions, pp. 121–2, New York: Wiley – Interscience 1973
274. Haworth, R.D., McKenna, J., Powell, R.G.: J. Chem. Soc. *1953*, 1110
275. Eliel, E.L. et al.: Conformational Analysis, p. 44, New York: Wiley – Interscience 1965
275a. van Eikeren, P.: J. Org. Chem. *45*, 4641 (1980)
276. Kirby, A.J., Martin, R.J.: J. Chem. Soc. Chem. Commun. *1978*, 803
277. Martin, R.J.: Thesis: Cambridge, 1979
278. Kirby, R.J., Martin, R.J.: J. Chem. Soc. Chem. Commun. *1979*, 1079
379. Evans, C.M., Glenn, R., Kirby, A.J.: unpublished work
280. Baldwin, J.: J. Chem. Soc. Chem. Commun. *1976*, 734 – 8
281. Baggett, N. et al.: Carbohydrate Research *2*, 216 (1966). Aksnes, G., Albriktsen, P., Juvvik, P.: Acta Chem. Scand. *19*, 920 (1965). Aksnes, G., Albriktsen, P.: Acta Chem. Scand. *20*, 1330 (166)
282. Reddy, C.P., Singh, S.M., Rao, R.B.: Tetrahedron Letters *1981*, 973

283. Craze, G.-A., Kirby, A.J.: J. Chem. Soc. Perkin 2 *1978*, 357
284. Knier, B., Jencks, W.P.: J. Am. Chem. Soc. *102*, 6789 (1980)
285. Jencks, W.P.: Accts. Chem. Res. *13*, 161 (1980)
286. Fife, T.H.: J. Am. Chem. Soc. *89*, 3228 (1967). Capon, B., Page, M.: J. Chem. Soc. Chem. Commun. *1976*, 1443
287. Salomaa, P.: Acta Chem. Scand. *20*, 1263 (1966). Schaleger, L.L., Richards, C.N.: J. Am. Chem. Soc. *92*, 5565 (1970)
288. Capon, B., Page, M.I.: J. Chem. Soc. Perkin 2 *1972*, 522
289. Fife, T.H., Hagopian, L.: J. Am. Chem. Soc. *90*, 1007 (1968)
290. Beaulieu, N., Dickinson, R.A., Deslongchamps, P.: Can. J. Chem. *58*, 2531 (1980)
291. Deslongchamps, P., Rowan, D.D., Pothier, N.: Can. J. Chem. *59*, 2787 (1981)
292. Swain, C.G. et al.: J. Am. Chem. Soc. *101*, 3576 (1979)
293. Wheeler, O.H.: in The Chemistry of the Carbonyl Group (ed.) Patai, S., vol. 1, p. 529, New York: Interscience 1966
294. Perlin, S.: Can. J. Chem. *49*, 1972 (1971)
295. Angyal, S.J., James, K.: Carbohydrate Research *12*, 147 (1970)
296. Angyal, S.J., James, K.: Aust. J. Chem. *23*, 1209 (1970)
297. Bazbouz, A. et al.: Tetrahedron *36*, 2757 (1980). McClelland, C.W.: J. Chem. Soc. Chem. Commun. *1979*, 751. Hanessian, S., Plessas, N.R.: J. Org. Chem. *34*,
298. Steinbeck, K.: J. Chem. Res. (S) *1980*, 94
299. Meyers, A.I. et al.: Tetrahedron Letters *1979*, 4159
300. Eliel, E.L., Hartmann, A.A., Abatjoglou, A.G.: J. Am. Chem. Soc. *96*, 1807 (1974)
301. Deslongchamps, P., Moreau, C.: Can. J. Chem. *49*, 2465 (1971)
302. Deslongchamps, P. et al.: Can. J. Chem. *50*, 3402 (1972)
303. Deslongchamps, P. et al.: Can. J. Chem. *52*, 3651 (1974)
304. Deslongchamps, P. et al.: Can. J. Chem. *53*, 1204 (1975)
305. Taillefer, R.J.: Can. J. Chem. *57*, 3040 (1979)
306. Kovać, F., Plesnica, B.: J. Am. Chem. Soc. *101*, 2677 (1979)
307. Deslongchamps, P. et al.: Can. J. Chem. *50*, 3405 (1972): *53*, 1601 (1975)
307a. Beaulieu, N., Deslongchamps, P.: Can. J. Chem. *58*, 164 (1980)
307b. McClelland, R.A., Alibhai, M.: Can. J. Chem. *59*, 1169 (1981)
308. Deslongchamps, P. et al.: Can. J. Chem. *58*, 1051 (1980)
309. Capon, B., Ghosh, A.K., Grieve, D.McL.A.: Accts. Chem. Res. *14*, 3061 (1981)
310. Ahmad, M. et al.: J. Am. Chem. Soc. *99*, 4827 (1977): *101*, 2669 (1979)
311. King, J.F., Allbutt, A.D.: Can. J. Chem. *48*, 1754 (1970)
312. Khouri, F., Kaloustian, M.K.: J. Am. Chem. Soc. *101*, 2249 (1979): Tetrahedron Letters *1978*, 5067
313. Kaloustain, M.K., Khouri, F.: J. Am. Chem. Soc. *102*, 7579 (1980)
314. Bender, M.L.: J. Am. Chem. Soc. *75*, 5986 (1953). Moffatt, A., Hunt, H.: J. Am. Chem. Soc. *81*, 2082 (1959)
315. Bladon, P., Forrest, G.C.: J. Chem. Soc. Chem. Commun. *1966*, 481
316. Hine, J., Ricard, D., Perz, R.: J. Org. Chem. *38*, 110 (1973)
317. Woodward, R.B.: Pure Appl. Chem. *9*, 49 (1964). Woodward, R.B., Gougoutas, J.Z.: J. Am. Chem. Soc. *86*, 5030 (1964). Goto, T. et al.: Tetrahedron *21*, 2029 (1965)
318. Kirby, A.J.: in Comprehensive Chemical Kinetics (ed.) Bamford, C.H., Tipper, C.F.H., vol. 10, p. 57, Amsterdam: Elsevier 1972
319. Deslongchamps, P.: Nouveau J. de Chim. *1*, 235 (1977)
320. Silver, B.L., Samuel, D.: Adv. Phys. Org. Chem. *3*, 123 (1965)
321. Long, F.A., Friedmann, L.: J. Am. Chem Soc. *72*, 3692 (1950)
322. Pocker, Y., Green, E.: J. Am. Chem. Soc. *95*, 113 (1973)
323. Bunton, C.A., Nayah, B., O'Connor, C.: J. Org. Chem. *33*, 572 (1968)

324. Deslongchamps, P. et al.: Nouveau J. de Chim. *2*, 631 (1978)
325. Deslongchamps, P., Taillefer, R.J.: Can. J. Chem. *58*, 2167 (1980)
326. Guthrie, J.P.: J. Am. Chem. Soc. *96*, 3608 (1974)
327. Caswell, M., Schmir, G.L.: J. Am. Chem. Soc. *101*, 7323 (1979). Smith, V.F., Schmir, G.L.: J. Am. Chem. Soc. *97*, 3171 (1975)
327a. McClelland, R.A., Potter, J.P.: Can. J. Chem. *58*, 2318 (1980)
328. Deslongchamps, P.: Nouveau J. de Chim. *3*, 343 (1979)
329a. Halliday, J.D., Symons, E.A.: Can. J. Chem. *56*, 1463 (1978)
329b. Burdick, B.A., Benkovic, P.A., Benkovic, S.J.: J. Am. Chem. Soc. *99*, 5716 (1977)
330. Deslongchamps, P., Taillefer, R.J.: Can. J. Chem. *53*, 2791, 3029 (1975)
331. Allen, P., Ginos, J.: J. Org. Chem. *28*. 2759 (1963)
332. Deslongchamps, P., Cheriyan, U.O., Taillefer, R.J.: Can. J. Chem. *57*, 3262 (1979)
333. Deslongchamps, P., Caron, M.: Can. J. Chem. *58*, 2061 (1980)
334. Krabbenhoft, H.O., Wiseman, J.R., Quinn, C.B., J. Am. Chem. Soc. *96*, 258 (1974)
335. Meyer, W.P., Martin, J.C.: J. Am. Chem. Soc. *98*, 1231 (1976)
335a. Neumann, R.C., Jones, V.: J. Org. Chem. *39*, 929 (1974)
336a. Johnson, J.E., Cornell, S.C.: J. Org. Chem. *45*, 4144 (1980): *46*, 3623 (1981)
336b. Hegarty, A.F., McCarthy, D.G.: J. Am. Chem. Soc. *102*, 4537 (1980). Hegarty, A.F.: Accts. Chem. Res. *13*, 448 (1980)
337. Lehn, J.-M., Wipff, G.: J. Chem. Soc. Chem. Commun. *1975*, 800
338. Gorenstein, D.G. et al.: J. Am. Chem. Soc. *99*, 4170 (1977)
339. Gorenstein, D.G., Findlay, J.B.: J. Am. Chem. Soc. *99*, 8048 (1970)
340. Gorenstein, D.G., Luxon, B.A., Findlay, J.B.: J. Am. Chem. Soc. *101*, 5869 (1979)
341. Gorenstein, D.G., Luxon, B.A., Goldfield, E.M.: J. Am. Chem. Soc. *102*, 1757 (1980)
342. Lehn, J.-M., Wipff, G.: J. Am. Chem. Soc. *96*, 4048 (1974)
343. Rowell, R., Gorenstein, D.G.: J. Am. Chem. Soc. *103*, 5894 (1981)
344. Hall, C.R., Inch, T.D.: J. Chem. Soc. Perkin 1 *1981*, 2368. Tetrahedron *36*, 2059 (1980)
345. Corriu, R.J.P. et al.: Tetrahedron *35*, 2889 (1979)
346. Gehrke, S.H., Wadsworth, W.S.: J. Org. Chem. *45*, 3921 (1980)
347. Font Freide, J.J.H.M., Trippett, S.: J. Chem. Soc. Chem. Commun. *1980*, 157
348. Brown, C. et al.: J. Chem. Soc. Perkin 2 *1976*, 888
349. Gregory, A.R., Malatesta, V.: J. Org. Chem. *45*, 122 (1980)
350. Malatesta, V., Ingold, K.U.: J. Am. Chem. Soc. *103*, 609 (1981)
351. Griller, D. et al.: J. Am. Chem. Soc. *103*, 619 (1981)
352. Beckwith, A.L.J., Easton, C.J.: J. Am. Chem. Soc. *103*, 615 (1981)
353. Hayday, K., McKelvey, R.D.: J. Org. Chem. *41*, 2222 (1976)
354. Rynard, C.M., Thankachan, C., Tidwell, T.T.: J. Am. Chem. Soc. *101*, 1196 (1979)
355. Gilbert, B.C., King, D.M., Thomas, C.B.: J. Chem. Soc. Perkin 2 *1980*, 1821
356. Burton, G.W. et al.: J. Am. Chem. Soc. *102*, 7791 (1980). Burton, G.W., Ingold, K.U.: J. Am. Chem. Soc. *103*, 6472 (1981)
357. Skell, P.S., May, D.D.: J. Am. Chem. Soc. *103*, 967 (1981)
358. Gorenstein, D.G. et al.: J. Am. Chem. Soc. *99*, 3473 (1977)
359. Imoto, T. et al. in The Enzymes (ed.) Boyer, P.D., vol. VII, p. 665, New York: Academic Press 1972
360. Gandour, R.D.: Bioorganic Chem. *10*, 169 (1981)

F. Subject Index

A-values 5, 7
A1, A2, A3-conformations 18–19, 51
Acetals,
  bonding in 67–71
  bond lengths and angles 53–61
  conformation 20, 49, 68–71, 99–100
  formation 94
  hydrolysis 84–94
  oxidation 96–102, 109
Acetoxy-radical 132
2-Acetoxy tetrahydropyran 61
Acetylenes 14, 25
2-Alkoxytetrahydropyrans 18–20, 51, 57
Amides, ^{18}O-exchange 114–7
Amidine hydrolysis 119–20
Ammonia, electronic structure 40
Ammonium, electronic structure 40
Anomeric effect: 14–15, 39
  at sp^2 centres 72–75
  generalised 20, 71
  in acetylenes 14, 25
  in furanoses 61
  in N-compounds 14
  in N-glycosides 15
  in nucleosides 61
  in P compounds 26–31
  in S compounds 23, 31
  of carboxy-group 25
  of alkoxy-group 9–10, 57
  of halogens 10–11
  of hydroxy-group 7–8
  origin 49, 65–70
  reverse 15–18, 27, 71, 85
  secondary 106–7
Aza-cyclohexanes, conformation 22–3

Boat (twist) conformations 12, 15–16, 30, 87, 128, 133
Bond angles 52–62, 69–71
Bond lengths: 39, 52–62
  and reactivity 90–1

Calculations 48–52
CCl_4, CF_4 52
1-Chloro-9-azabicyclo [3.3.1] nonane 125–6
Chloroformate esters 75
1-Chloro-9-oxabicyclo [3.3.1] nonane 125–6
CH_2Cl_2 52
CH_2F_2 52, 62–64
$CH_2=\overset{+}{O}H$ 82
$CH_2=\overset{+}{O}Me$ 82
Conformation:
  A-values 5, 7
  and reactivity 82–84
  of acetals 20, 99–100
  of 2-alkoxytetrahydropyrans 18–20 51
  of 2-alkylthiotetrahydropyrans 24
  of ethers 4
  of glucose 5–6
  of mannose 7
  of orthoesters 103–5
  of pyranoses 5, 8, 10, 12, 19
  of tetrahydropyrans 4–14
  of thiacetals 25–6
Crystal structures 19–20, 43–48, 56–61, 90
Cyclic AMP 36

Deformation density 45
1,3-Diaxial interactions 5, 12–14
  between OAcyl groups 14
  in azacyclohexanes 23
  in 1,3-dioxanes 21
  in S-heterocycles 23–24

2,3-Dichloro-1,4-dioxanes 24, 38–9, 53
2,3-Dichloro-1,4-dithianes 24
2,3-Dichloro-1,4-thioxanes 24
1,2-Difluoroethane 32–4, 77
Dihydroxymethane 49–50, 53, 67–8
Dimethoxymethane 20, 49–52, 54–56, 67–71
Dimethyl phosphate 55–56, 126–7
1-(2,4-Dinitrophenoxy)-9-oxabicyclo [3.3.1] nonane 89–90
1,8-Dioxadecalin 100
1,3-Dioxanes
  conformation 21
  oxidation by O_3 100–1
1,3-Dioxolanes, hydrolysis 91–93
Dipole-dipole interactions 21, 38
1,3-Dithianes 23, 34

E1cb reaction 80–1
E1, E2, E3 conformations 18–19, 51
E2 reaction 86–7
Electron density 42, 44–8
Enzyme reactions 132–5
Esters:
  anomeric effect in 73, 106
  carbonyl ^{18}O-exchange 112–3
Exo-anomeric effect 18–20, 24–6, 72

$FCH_2CH_2^-$ 80
Fluoroformate esters 75
Fluoromethanol 33, 48, 66–7
FOOF 75–6
Furanoses 61

Gauche effect: 32–36, 75–77
  attractive 35–6
  repulsive 35
β-Glycosidases 133
Glycopyranosides:
  acetyl 13, 57, 70
  bond lengths 52
  electronegativity of ring-O 61
  glycosyl transfer reactions 133
  methyl 11, 13, 85–6, 97, 99–100
Glycopyranosyl halides 58–9

5-Halogeno-1,3-dioxanes 34
α-Halogeno-ethers 33–4, 38, 53–5, 64, 70, 80

Hockey sticks effect 35
Hofmann elimination 86–7
H_2S_2 32
Hybridisation 39–42, 78–9, 135–6
Hydrazines 33
H-bonding 5, 42
  directionality 43-4, 47
Hydroxyethylene 73
Hydroxy group:
  A-value 5, 7–8
  size 5
  solvation 5, 8

Imidate esters 117–126
Imidoyl chlorides 73, 126

Lactams, ^{18}O-exchange 114–5
Lactones, ^{18}O-exchange 113
Lone pair electrons: 37–8
  definition 40–1
  directionality 41, 43
  molecular orbitals 135–7
  physical reality 41–8
  size 41
Lysozyme 133–4

Methoxy group, anomeric effect 9
1-Methoxyethanol 54
Methoxymethanol 54
Methoxymethyl chloride 23–4, 67
Methoxymethyl fluoride 49–50, 66–7
Methoxymethyl formate 49–50, 55

N-glycosides:
  anomeric effect in 14
  barriers to rotation in 80
  reverse anomeric effect in 15–17
2-(4-nitrophenoxy)-**trans**-1-oxadecalins 86–8
n-π^* interactions 63–6, 69–71
n-σ^* interactions 39, 64–80, 133, 135
Nucleosides 61
Nuclear quadrupole moment 80

Orthoesters:
  conformation 103–5
  hemiorthoesters 105–110

hydrolysis 102–112
Oxygen exchange:
amides 114
esters 112

Pentaoxyphosphoranes 93, 127
Phosphate esters:
conformation 55–56, 134
cyclic 28, 127–8
diester hydrolysis 134
pentacovalent 126–8
Phosphonate esters, cyclic 29
Phosphoramidate esters, cyclic 30

Rabbit-ear effect 38
Radicals:
reactions 128–132
stabilisation by α-O 129
Ribonuclease A 134

S-Methylthianium 31
σ-σ^* interactions 77
Stereoelectronic control:
conditions for 83
of acetal formation 94–6
of acetal hydrolysis 88–93
of enzyme reactions 132
of glycosyl transfer 133
of imidate ester hydrolysis 117–126
of orthoester hydrolysis 102–112
of radical reactions 128–132
Stereoelectronic effects at N 117–126
S_N2 reaction 79
Sulphate esters, conformation 32
Sulphite esters, cyclic 31
Sulphonate esters, conformation 32
Sulphoxides 31

Tetrahedral intermediates 113–125
Tetrahydrofuran-2-yl radical 128
Tetrahydropyran-2-yl radical 128
Thianes: 2-alkoxy 25
2-phenylthio 25
Thioacetals, conformation 25–6
tRNA 134
1,3,5-Trithianes 23, 25

Uridines, 5-substituted 62

Vitamin E 131